鮮美魚料理100道

一學就

程安琪 著

Angela Cheng

作　　者　程安琪

發 行 人　程安琪
總 策 畫　程顯灝
編輯顧問　錢嘉琪
編輯顧問　潘秉新

總 編 輯　呂增娣
主　　編　李瓊絲
主　　編　鍾若琦
執行編輯　許雅眉
編　　輯　吳孟蓉、程郁庭
美術主編　潘大智
美　　編　李傳慧
行銷企劃　謝儀方
出 版 者　橘子文化事業有限公司

總 代 理　三友圖書有限公司
地　　址　106台北市安和路2段213號4樓
電　　話　(02) 2377-4155
傳　　真　(02) 2377-4355
E－mail　service@sanyau.com.tw
郵政劃撥　05844889 三友圖書有限公司

總 經 銷　大和書報圖書股份有限公司
地　　址　新北市新莊區五工五路2號
電　　話　(02) 8990-2588
傳　　真　(02) 2299-7900

初　　版　2014年2月
定　　價　新臺幣299元
ISBN　978-986-6062-85-8（平裝）

國家圖書館出版品預行編目(CIP)資料

鮮美魚料理100道，一學就會，輕鬆上桌／程安琪作. -- 初版. -- 臺北市：橘子文化，2014.02 面；　公分
ISBN 978-986-6062-85-8(平裝)

1. 海鮮食譜 2. 魚

427.252　　103000328

序

「多吃魚才會聰明」，從小就聽媽媽這麼說，同時她還會說出許多魚的好處，哄我們多吃魚。除了是魚的營養價值真的很好之外，當然也是因為媽媽新研究、尚在實驗階段的魚料理，需要我們品嚐、捧場。也可想而知，在我們家餐桌上出現的魚，就不是一般紅燒或乾煎一類簡單的吃法了！

從小養成吃魚的習慣，同時魚又可以有許多不同的變化，使我非常愛吃魚。每次走到魚攤前，就忍不住停下來看看，腦中也出現許多魚燒好的畫面，勾起我的嘴饞。遇到旁邊買魚的人問老闆，「這條魚該怎麼煮呢？」也忍不住要插嘴說一下心得。想想為什麼不出一本教大家怎麼烹調出好吃的魚的食譜呢？

有些魚是我們常吃又很大眾化的魚，另外魚攤上也會出現一些叫不出名字的魚，其實魚的名字並不重要，重要的是要知道魚肉的屬性，找尋適合它的烹調方法，才能引出完美的滋味。每一種烹調方法，無論蒸、炒、燒、炸、燙，其實都有它的訣竅。會做了，操作起來得心應手；不熟練、拿捏不到重點，簡單的煎魚也煎得支離破碎，蒸怕蒸不熟，燒又怕燒不入味，不知該如何下手烹調。

台灣近些年魚類的養殖非常進步，許多魚都物美價廉，而海水魚也因冷凍和冷藏技術的進步，選擇非常多。就以潮鯛魚片來說，大賣場中賣 1 片 250 公克，還不到 70 元，有些人嫌它略有土味，但是正確的處理後就能使它又嫩又好吃，食譜中將會教你如何處理。在烹調魚之前，選購和清洗都是很重要的，會挑選魚也要會清理。魚鱗要刮乾淨是第一步，大骨中殘留的污血要清乾淨，才不會使魚燒好後有腥氣和土味。

在本書中我為讀者們挑選了 100 種魚的烹調法，其實每一種口味並不限定用哪一種魚才好吃，像有名的「廣東式」蒸魚，我就用了馬頭魚，而不是一般認為的石斑或青衣，其實只要是新鮮的魚都可以用同樣方法來蒸。在「蒸」魚中有 13 種不同口味的蒸魚食譜，如果每種口味配上 8 種不同的魚，不就可以再變出 100 道魚的料理嗎？

最後也是最重要的是，在動手烹調之前，希望大家先把每一種烹調方法的重點先看一下，會使你更輕鬆地做出好吃的魚。

程安琪

目　錄
Contents

蒸

炒

燒

炸

烤

冷食

湯

Steam 蒸

在「蒸」之前——

「蒸」一直被認為是最能品嚐魚的鮮美滋味的最好方法，在廣東餐廳中，「清蒸游水魚」也是老饕的最愛。

廣東人特愛石斑魚，在香港，老鼠斑是常居排行榜第一名的魚種。另外「蘇眉」、「石鯛」、「梳齒」也都是適合清蒸的高價魚。其實只要魚夠新鮮，就是適合清蒸的魚。一般說來，肉質細嫩、緊實有彈性、沒有特殊味道的魚都可以蒸。淡水魚中的活魚，新鮮度當然不用說，海魚或是已殺好的淡水魚則要從魚眼、魚腮、魚鱗等部位來判斷，選一條新鮮的魚是蒸魚的第一步。

第二點要注意的就是火候，蒸魚要等水滾了才能放進去蒸，即使用電鍋蒸魚，也要先按下開關，等水蒸氣上來了，才能放進去蒸。同時最好多放 1 ～ 2 杯水，使水蒸氣充足一些才好。

放進蒸鍋後開始計時，用一般家庭火力來蒸整條魚，若魚是 12 兩（450 公克）時大約要蒸 11 分鐘，但是因為魚的種類不同，魚肉有厚薄的差別，每家的火力也有大小，不能一概而論，一定要用一支細筷子由魚頭下方、魚大骨的旁邊、魚肉最厚的地方插進去試一下，要能穿入、而取出的筷子上又沒有沾黏魚肉才是熟了。

掌握了新鮮度和火候，應該有及格的 60 分了！接著就是味道，廣東人只是用豉油皇（上好的醬油）和蔥絲、香菜搭配，就降服了許多人的胃。其實蒸魚還可以有許多變化，在後面的食譜中，其實每一種都可以互相搭配，例如喜歡豆豉味道的人，不只可以蒸草魚段，任何買到的新鮮魚都可以蒸。蘿蔔乾蒸魚也是值得嚐試的好滋味。

常有人問，蒸魚後的汁要倒掉嗎？一般如果沒有加料蒸，或魚汁腥氣較重，尤其是廣東式的蒸魚，都會倒掉，再另加調味汁。至於蒸的魚要不要先抹一點鹽、淋一點酒，也各有擁護者，加的話是可以使肉質更緊實一些，蒸的時候不會有太多汁液滲出。

至於用微波爐蒸魚，除了要注意時間外，另外在盤子裡多加 4 ～ 5 大匙的水以保持魚肉的嫩度，也是要注意的。

「蒸」一條美味的魚應該是很容易的事了吧！

一般推薦可用於蒸的魚種有

馬頭魚 ・ 紅新娘 ・ 草魚頭 ・ 黑格 ・ 鱈魚 ・ 金目鱸魚 ・ 梳齒 ・ 青衣 ・ 石雕

蘿蔔乾蒸魚頭
Steamed Fish Head with Dried Radish

材料
草魚頭 1/2 個（約 450 公克）
蔥花 1/2 杯、蘿蔔乾 150 公克

調味料
Ⓐ鹽 1/2 茶匙、酒 1 大匙
Ⓑ醬油 1 大匙、糖 2 茶匙
油 1½ 大匙

做法
1. 魚頭洗淨，抹上調味料Ⓐ放置 5 分鐘後放在蒸盤上。
2. 蘿蔔乾洗淨剁成細末（若蘿蔔乾太鹹時，要先泡水漂去鹹味），拌上調味料Ⓑ，撒在魚頭上。
3. 蒸鍋水滾後，放入魚頭蒸 20 ～ 25 分鐘，撒下蔥花，再蒸 1 ～ 2 分鐘即可。

Ingredients
1/2 fish head(about 450g), 150g dried radish , 1/2cup chopped green Onion

Seasonings
Ⓐ1/2t. salt, 1T. wine
Ⓑ1T. soy sauce, 2t. sugar, 1½T. oil

Procedure
1. Rinse the fish headⒶmarinate with seasoningsAfor 5 minutes, place on a plate.
2. Rinse and chop dried radish to very fine(soak in cold water to remove the salty taste first , if it taste to salty). Mix withⒷ, sprinkle on top of fish head.
3. Steam for 20~25 minutes after water boiled, sprinkle green onion on and steam for another 1~2 minutes. Serve hot.

豆豉蒸草魚段
Steamed Fish with Black Bean Sauce

材料
草魚 1 段（約 450 公克）、黑豆豉 1½ 大匙
薑屑 1 大匙、紅辣椒屑 1 大匙、蔥粒 1 大匙

調味料
Ⓐ蔥 2 支、薑 2 片、鹽 1/3 茶匙、酒 1 大匙
Ⓑ油 1 大匙、醬油 2 茶匙、糖 1 茶匙
水 2 大匙

做法
1. 草魚洗淨、擦乾。在表面劃上兩條刀口或斜切成片，用調味料Ⓐ抹擦，醃 10 分鐘。清除蔥、薑，放入蒸盤中。
2. 豆豉泡水約 5 分鐘，取出放在小碗中，加薑屑、紅辣椒屑及調味料Ⓑ調勻，淋在魚身上。入鍋中以大火蒸約 12 分鐘至熟。
3. 撒下蔥粒，蔥粒上淋下 1 大匙燒熱的油便可上桌。

Ingredients
450g fish(center portion), 1½T. fermented black beans, 1T. chopped ginger, 1 T. chopped red chili, 1T. chopped green onions

Seasonings
Ⓐ2 stalks green onion, 2 slices ginger, 1/3 t. salt
1 T. wine
Ⓑ1T. oil, 2t. soy sauce, 1t. sugar, 2T. water

Procedure
1. Rinse fish and pat dry. Score 2 times on both side or cut into two pieces. Marinate withⒶ for 3~5minutes. Discard green onion and ginger, place fish on a plate.
2. Soak the fermented black beans for 5 minutes, drain. Mix the beans with ginger, red chili and seasonings Ⓑ in a small bowl, drizzle over fish, steam over high heat for about 12 minutes until done.
3. Sprinkle green onion on fish, pour 1T. heated oil over fish to enhance the flavor.

樹子蒸鮮魚

Steamed Fish with Tree Seeds

材料

紅新娘魚 2 條（約 300 公克）
罐頭樹子（破布子）2 大匙
醬瓜 2 ～ 3 條、薑絲 1 大匙
蔥絲 1 大匙

調味料

醬瓜湯汁 2 大匙、樹子湯汁 1 大匙
酒 1/2 大匙

做法

1. 魚打理乾淨，放在抹了少許油的蒸盤上面。
2. 醬瓜切成和樹子差不多大小的丁，和樹子混合撒在魚身上，再撒下薑絲和調勻的調味料。
3. 入蒸鍋蒸至魚熟即可，撒下蔥絲即關火，取出上桌。

Ingredients

2 fresh fish(about 300g.), 2T. tree seeds(in can), 2~3 T. pickled cucumber, 1T. shredded ginger, 1T. shredded green onion

Seasonings

2T. juice from can of pickled cucumber, 1T. juice from can of tree seeds, 1/2T. wine

Procedure

1. Rinse fish, place on a greased plate.
2. Dice pickled cucumber into the size of tree seeds, mix them and sprinkle on top of fish, put ginger on and then pour over the mixed seasonings .
3. Steam fish until done, sprinkle green onion and turn off the heat. Serve.

蒜蓉蒸魚片
Steamed Fish Slices with Garlic Sauce

材料
紅尼羅魚 1 條或白色魚肉 300 公克
大蒜泥 1 大匙、蔥花 1 大匙

調味料
Ⓐ酒 2 茶匙、醬油 2 茶匙、鹽 1/4 茶匙
水 3 大匙
Ⓑ淡色醬油 1 大匙、胡椒粉少許
水 2 大匙、油 1 大匙

做法
1. 魚洗淨擦乾後由背部劃開，取下兩面的魚肉，打斜切片，排入盤中。
2. 大蒜泥和調味料Ⓐ調勻，淋在魚片上，入鍋蒸 5 分鐘。取出撒上蔥花。
3. 調味料Ⓑ在小鍋中煮滾，淋在蔥花上，同時將汁澆淋在魚片上。

Ingredients
1fish or 300g. fish fillet , 1T. smashed garlic, 1T. chopped green onion

Seasonings
Ⓐ2t. wine, 2t. soy sauce, 1/4t. salt, 3T. water
Ⓑ1T. light color soy sauce, a pinch of pepper, 2T. water, 1T. oil

Procedure
1. Rinse fish, remove two pieces of meat(with skin)from both sides, slice into big pieces, arrange on a plate.
2. Mix garlic and seasoningsⒶ well, drizzle over fish, steam for 5minutes. Remove, sprinkle green onion on top.
3. Bring seasoningsⒷ to a boil, pour on top of green onion, drizzle fish with the sauce.

芙蓉蒸魚球

Steamed Egg with Cod

材料

鱈魚 1 片（約 300 公克）、蛋 4 個、芹菜粒 1 大匙、香菜末適量

調味料

Ⓐ鹽 1/4 茶匙、太白粉 1/2 大匙
Ⓑ鹽 1/2 茶匙、水 1½ 杯
Ⓒ清湯或水 3/4 杯、醬油 2 茶匙、白胡椒粉少許、麻油少許

做法

1. 鱈魚去骨，將肉切成四方塊，加調味料Ⓐ拌勻，醃 5 分鐘後用滾水川燙 10 秒鐘。
2. 蛋加調味料Ⓑ打散，盛入水盤中，入鍋蒸至 8 分熟（先以大火蒸 3 分鐘，改小火再蒸約 10 分鐘），放上魚塊後，再蒸 4 ～ 5 分鐘。
3. 煮滾調味料Ⓒ，關火後放入芹菜粒，淋在魚球上，再撒上香菜末即完成。

Ingredients

1slice of cod(about 300g.), 4 eggs, 1T. chopped celery, a little of cilantro

Seasonings

Ⓐ1/4t. salt, 1/2T.cornstarch
Ⓑ1/2t. salt, 1½C. water
Ⓒ3/4C. soup stock or water, 2t. soy sauce, a pinch of pepper, a little of sesame oil

Procedure

1. Remove bones from cod, cut the meat into pieces. Marinate withⒶ for 5minutes. Blanch for 10 seconds, drain.
2. Beat eggs well with Ⓑ, pour into a deep plate, steam over high heat for 3 minutes, reduce to low heat for about 10minutes, until the egg get 80% firmed, place fish on top of egg, steam for another 4~5 minutes.
3. BringⒸ to a boil, turn off the heat and add celery, pour over fish, sprinkle cilantro sections.

豆酥魚片

Steamed Fish Filet with Bean Sauce

材料

白色魚肉 250 公克、百頁豆腐 1 條、黃豆豉 1/2 球、大蒜屑 1/2 大匙
薑末 1 茶匙、蔥末 2 大匙

調味料

Ⓐ鹽、酒、胡椒粉各少許
Ⓑ辣豆瓣醬 1/2 茶匙、酒 1 茶匙、糖 1/4 茶匙、麻油少許

做法

1. 百頁豆腐切厚片，用滾水快速燙一下，瀝乾水分放在盤子上。
2. 魚片打斜切片，拌上調味料Ⓐ，排在豆腐上，蒸約 7 ～ 8 分鐘至熟，取出倒掉湯汁。
3. 黃豆豉剁得非常細。鍋中用 3 大匙油先炒蒜屑、薑末和豆豉屑，小火炒出香味且成為金黃色後，加入辣豆瓣醬等調味料Ⓑ炒勻。
4. 趁熱淋在魚片上，撒下蔥末即可。

Ingredients

250g. fish fillet, 1 tender dried bean curd, 1/2 ball of fermented soy beans, 1/2T. chopped garlic, 1t. chopped ginger, 2T. chopped green onion

Seasonings

Ⓐa little of salt, wine and pepper
Ⓑ1/2t. hot soy bean paste, 1t. wine, 1/4t. sugar, 1/4t. sesame oil

Procedure

1. Cut the bean curd into thick pieces, blanch and drain, arrange on a plate.
2. Slice fish, mix withⒶ, place on top of bean curd, steam for 7~8 minutes. Discard the liquid.
3. Chop the fermented soy beans until very fine. Heat 3T. oil to stir~fry ginger and garlic and soy beans until very dry and crispy. Add seasoningsⒷ, mix well.
4. Drizzle over fish, sprinkle green onion, serve.

廣式清蒸魚

Steamed Fish , Cantonese Style

材料

新鮮魚 1 條（約 450 公克）、蔥 2 支
薑絲 2 湯匙、蔥絲 1/2 杯、香菜段 1/2 杯

調味料

醬油 2 大匙、糖 1/2 茶匙、水 3 大匙
白胡椒粉少許

做法

1. 魚打理乾淨，擦乾水分。兩面均劃上刀口，蔥切成長段。
2. 在盤子上墊上蔥段，放上魚後撒上薑絲，入蒸籠內用大火蒸約 10 ～ 11 分鐘，用竹筷試插魚肉，確定已熟後端出。倒出蒸魚汁，夾掉蔥段，撒下白胡椒粉。
3. 炒鍋中燒熱 2 大匙油，淋下調勻的調味料，一滾即關火，撒下蔥絲和香菜段，全部淋在魚身上。

Ingredients

1 fresh fish(about 450g.), 2 stalks green onion, 2T. shredded ginger, 1/2C. shredded green onion, 1/2C. cilantro sections

Seasonings

2T. soy sauce, 1/2t. sugar, 3T. water, a pinch of white pepper

Procedure

1. Rinse fish, pat dry. Scored 2 or3 times on each side. Cut green onion into long sections, place on the steaming plate.
2. Put fish on, sprinkle ginger over fish, steam with high heat for about 10~11minutes, try with a chopstick to make sure that the fish is done. Pour away the liquid from steaming fish, discard green onion, sprinkle pepper on.
3. Heat 2T. oil, pour mixed seasonings in, turn off the heat when it boils, put green onion shreds and cilantro sections in sauce, pour over fish.

XO 醬蒸魚球
Steamed Fish Cubes with XO Sauce

材料
白色魚肉 300 公克、西洋菜 1 把
蔥花 1 大匙、太白粉 1 大匙

調味料
Ⓐ麻油 1 茶匙、鹽 1/4 茶匙
ⒷXO 醬 2 大匙、鹽 1/4 茶匙、水 2 大匙
辣油 1/2 茶匙

做法
1. 西洋菜切約 3 公分段，用滾水燙軟，沖涼並擠乾水分，拌上調味料Ⓐ，放在盤中。
2. 魚肉切塊，拌上約 1 大匙的太白粉，放在西洋菜上面，再淋下已調好的調味料Ⓑ，蒸約 8 分鐘。
3. 開鍋趁熱撒上蔥花，取出上桌。

Ingredients
300g. fish fillet, 1 bundle of water cress, 1T. chopped green onion, 1T. cornstarch

Seasonings
Ⓐ1t. sesame oil, 1/4t. salt
Ⓑ2T. XO sauce, 1/4t. salt, 2T. water, 1t. chili oil

Procedure
1. Cut water cress into 3cm sections, blanch and rinse cold. Mix withⒶ, place on the steaming plate.
2. Cut fish into big cubes, mix with 1T. cornstarch, place on top of water cress, drizzle the mixed Ⓑ, steam for about 8 minutes.
3. Turn the heat, sprinkle green onion, serve hot.

雙冬蒸鮮魚
Steamed Fish with Black Mushroom Sauce

材料

新鮮魚 1 條（約 450 公克）、蔥花 1 大匙
香菇 2 朵、冬菜 2 大匙、蔥絲 1 大匙

調味料

醬油 1 大匙、糖 2 茶匙、水 1/3 杯

做法

1. 魚打理乾淨，擦乾水分，在肉厚處劃上刀痕，放在蒸盤上。
2. 香菇泡軟、切成細絲。冬菜泡水 1～2 分鐘，沖淨沙子，擠乾水分。
3. 起油鍋用 2 大匙油炒香蔥花、香菇和冬菜，加入調味料，煮滾後淋在魚上。
4. 入蒸鍋大火蒸 15 分鐘，再撒上蔥絲即可。

Ingredients

1 fresh fish(about 450g.), 1T. chopped green onion ,2 black mushrooms, 2T. salted cabbage, 1T. shredded green onion

Seasonings

1T. soy sauce, 2t. sugar, 1/3C. water

Procedure

1. Rinse fish and pat it dry. Make 2 cuts on the meat if it is too thick, place on a plate.
2. Soak and shred the black mushrooms. Soak salted cabbage with water for 1~2 minutes, rinse away sands, squeeze dry.
3. Heat 2T. oil to stir~fry green onion, black mushroom and salted cabbage, add seasonings. Bring to a boil, pour over fish.
4. Steam over boiling water with high heat for about 15 minutes, put green onion shreds on top, serve.

酸辣屑子魚

Steamed Fish with Spicy Meat Sauce

材料

鯧魚 1 條（約 12 兩）、絞肉 3 大匙、蔥屑 2 大匙
木耳屑 2 大匙、大蒜屑 1 大匙、薑屑 1 茶匙
芹菜屑 2 大匙、紅辣椒屑 1 大匙

調味料

醬油 2 大匙、酒 1 大匙、水 4 大匙、鹽 1/2 茶匙
糖 1 茶匙、胡椒粉 1/4 茶匙、太白粉 2 茶匙
麻油 1 茶匙、醋 3 大匙

做法

1. 將鯧魚打理乾淨，在背部肉厚處直劃一刀，抹少許鹽和酒放置 5 分鐘，上鍋蒸熟（約 10 分鐘），倒出蒸魚所出的湯汁。
2. 鍋中用 1 大匙油炒散絞肉，再加大蒜屑、薑屑及木耳屑炒香，倒入調勻的調味料炒滾，撒下芹菜屑、紅辣椒屑及蔥屑，即完成酸辣屑子汁。
3. 將屑子汁淋在魚上便可上桌。

Ingredients

1 pomfret (about 450g.), 3T. ground pork, 2T. chopped black fungus, 1T. chopped garlic, 1t. chopped ginger, 2T. chopped celery, 1T. chopped chili, 2T. chopped green onion

Seasonings

2T. soy sauce, 1 T. wine,4 T. water, 1/2t. salt, 1t. sugar, 3T. vinegar, 1/4 t. pepper, 1t. sesame oil, 2t. cornstarch

Procedures

1. Rinse fish. Make 1~2 cuts on each side. Sprinkle some salt and wine on, leave for 5 minutes. Steam over high heat for about 10 minutes until done. Pour out the liquid from steaming fish.
2. Heat 1T. oil to stir-fry the ground pork, add garlic, ginger and fungus, add mixed seasonings, bring to a boil. Sprinkle celery, red chili and green onion at least. that's the spicy meat sauce.
3. Pour the meat sauce over fish, serve.

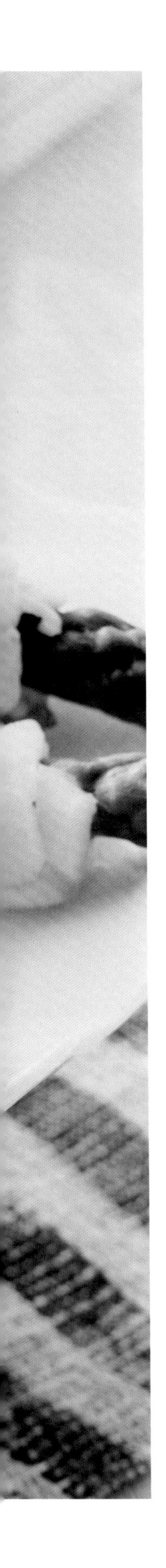

碧綠魚捲
Steamed Fish Rolls with Asparagus

材料
白色魚肉 450 公克、蘆筍 8 支、紅甜椒 1/4 個、蔥 2 支

調味料
Ⓐ鹽和胡椒粉各少許
Ⓑ蠔油 1/2 茶匙、酒 1 茶匙、鹽 1/4 茶匙、油 1/2 茶匙、麻油 1/2 茶匙
水 2/3 杯、太白粉 1/2 茶匙

做法
1. 魚肉打斜切成大片，排在砧版上，撒少許鹽和酒，拍勻，放置一下。
2. 將蘆筍削去硬皮，用熱水川燙 1 分鐘（水中加鹽少許），撈出後用冷水沖泡至涼。蔥切段。
3. 切下蘆筍嫩的尖端，包捲入魚片中。魚捲排在抹油的蒸盤中，放上蔥段，蒸 5 分鐘便可取出，換入餐盤中。
4. 其餘部分的蘆筍切長段，和切成粗絲的紅甜椒同炒一下，加少許鹽和胡椒粉調味，放置盤中。
5. 將調味料Ⓑ煮滾，淋在魚捲和炒蘆筍上。

Ingredients
450g.fish fillet, 8 stalks asparagus, 1/4 red bell pepper, 2stalks green onion

Seasonings
Ⓐa pinch of salt and pepper.
Ⓑ1/2t. oyster sauce,1t. wine, 1/4t. salt, 1/2t. oil, 1/2t. sesame oil, 2/3C. water, 1/2t. cornstarch

Procedure
1. Slice fish fillet into big pieces, sprinkle some salt and wine, sit aside for a few minutes.
2. Trim and discard the tough parts of asparagus. Boil in salted water for 1 minute, rinse to cool. Cut green onion into sections.
3. Cut the tips and roll it into fish meat. Arrange on a greased plate, put green onion on top, steam for about 5 minutes. Transfer to the serving plate.
4. Cut the rest part of asparagus into sections, stir-fry with red bell pepper, season with salt and pepper, place on serving plate.
5. Bring the seasoningsⒷ to a boil, pour over the fish rolls and asparagus.

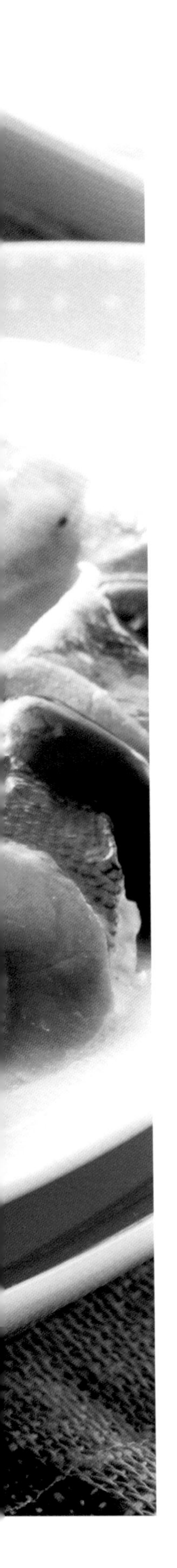

麒麟魚

Steamed Tri-color Fish

材料

鱸魚（或石斑魚）1 條（約 750 公克重）、香菇 6 朵、熟火腿 12 片
薑片 4 片、青花菜適量

調味料

Ⓐ鹽 1/2 茶匙、酒 1/2 大匙
Ⓑ清湯 1 杯、鹽少許、麻油 1 茶匙、太白粉 2 茶匙

做法

1. 將魚打理乾淨後切下頭和尾。由背部下刀沿著脊骨劃切，取下整片魚肉（兩面），剔除腹部的魚刺。
2. 將每一片魚肉切成 6 小片，用酒及鹽抹一下後，按原形在盤內排成 2 行（盤上抹少許油）。
3. 香菇泡軟切片。取香菇和火腿片各 1 片，夾在魚片中間，使成白黑紅相間。薑片散置在魚上面。用大火蒸約 7 分鐘即可取出。
4. 倒棄汁液，排放燙煮過之青花菜。小鍋內煮滾調味料Ⓑ，淋到魚上以增香氣及光亮。

Ingredients

750g. fresh fish, 6 black mushrooms, 12 pieces Chinese ham (cooked), 4 slices ginger, broccoli

Seasonings

Ⓐ1/2 t. salt, 1/2 T. wine
Ⓑ1C. soup stock, a little of salt, 1t. sesame oil, 2t. cornstarch

Procedure

1. Scale the fish. Remove two fish fillet from the body. Also remove all the small bones.
2. Slice fillet into 6 pieces from each side . Mix with wine and salt. Arrange on a greased plate in two rows.
3. Soak and slice each black mushroom into 2 pieces. Place one black mushroom and one ham between two pieces of fish meat. Put ginger slices on top and steam over high heat for 7minutes. Remove.
4. Discard ginger and liquid, place boiled broccoli on plate. Boil seasoningsⒷ in a sauce pan, pour over fish and broccoli, serve.

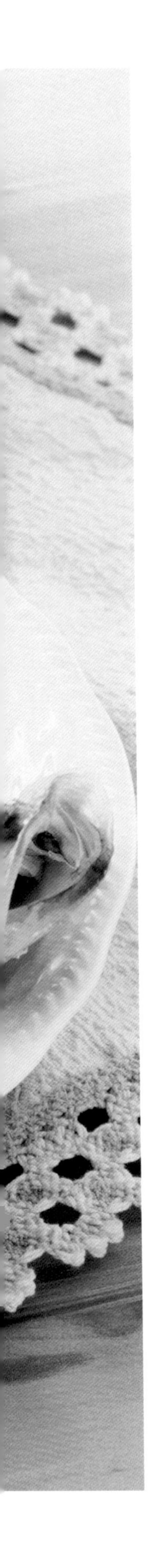

泰式檸檬魚
Lemon Fish , Tai Style

材料

鱸魚 1 條、大蒜末 1 大匙、紅辣椒末 2 茶匙、香菜末 1 大匙、香菜段適量

調味料

Ⓐ調味料鹽少許、酒 1 茶匙

Ⓑ糖 2 茶匙、檸檬汁 3 大匙、味露 2 大匙、熱高湯（或水）1 杯

做法

1. 鱸魚洗淨，由腹部剖開，但背部仍相連成一整片。放在蒸盤上，撒上調味料Ⓐ的鹽和酒。
2. 水滾後放魚盤入蒸鍋內，用大火蒸約 8 ～ 9 分鐘。見魚已熟，取出魚盤，將蒸魚汁倒掉。
3. 碗中將香菜末、蒜末、辣椒末和調味料Ⓑ調勻，淋到魚身上，再放入蒸籠內蒸約 30 秒後取出，撒上香菜段上桌。

Ingredients

1 sea perch, 1T. chopped garlic, 2t. chopped red chili, 1T. chopped cilantro, cilantro sections

Seasonings

Ⓐa pinch of salt, 1t. wine

Ⓑ2t. sugar, 3T. lemon juice, 2T. fish sauce, 1C. hot soup stock or water

Procedure

1. Rinse fish, cut open from the stomach side, but keep the back still connected. Place on a plate, sprinkle seasoningsⒶ.
2. When water boiled, steam over high heat for 8~9 minutes. Remove when the fish is done. Discard the liquid from fish.
3. Mix cilantro, garlic, red chili and seasoningsⒷ well, pour over fish, steam for another 1/2 minute, remove and sprinkle cilantro sections.

Stir fry 炒

在「炒」之前——

「炒」是中國菜中最常用、也是應用最廣的烹調方法，但是用來炒魚片、魚球的時候，就需要有些技巧了。

在炒魚片的時候，最好是經過「滑油」的動作，也稱做「盪鍋」，就是將炒鍋先燒熱，再把 1/2 杯的油倒入一起加熱。加熱時要搖轉鍋子，使油盪開，滋潤鍋子，然後再將油倒出。油鍋經過滑油的步驟，在炒或煎的時候才不會黏鍋，炒起來才能翻炒自如。

在選擇魚肉時，要選新鮮的魚肉，切片要順絲，不可切得太薄，而上漿之後，過油的溫度也應特別注意。太冷則不滑，太熱就會都黏成一團，同時小心不能太過翻動，以免魚肉碎散。

一般而言，新鮮的魚當然以活殺的最好，但價格較貴又不容易買到。從前都是以去骨去皮的石斑魚、大黃魚、鮸魚、草魚為主，現在多了現成冷凍的潮鯛魚片和大量養殖的鱸魚可選擇。其實較大型的鯧魚、青衣或紅尼羅魚等一些肉厚、又沒有小刺的魚，都可以取肉來炒（取肉後的魚頭、魚骨可以留做魚高湯，千萬不要丟掉）。我個人很喜歡用潮鯛魚片，方便又不貴，但出產的品牌也要挑選一下，以免會有較重的土腥味，另外就是要將魚片中紅色的肉修掉，紅色部分土味較重。同時切好片後，再用水沖洗一下。

要使炒過的魚片滑嫩，一定要上漿過。上漿醃的時候，鹽的量不要多，尤其是要醃較長的時間。另外如果是用潮鯛魚片，因為它的肉質較緊又經過冷凍，醃的時候最好先加水和鹽抓拌，使魚片吸水膨脹（一般魚肉則不用加水）。再拌上蛋白，並要持續抓拌，使蛋白吸收，最後再加太白粉拌勻。拌好後要放入冰箱冷藏，使蛋白和太白粉吸收、附著。這個給魚片加分的工作是不能省的。至於醃的時候加糖、薑汁、胡椒粉都是口味上的改變，可以自己取捨。注意的是，醃魚時是不放酒的，在炒的時候才烹酒增香。

如果怕過油再炒會吸取較多量的油份，也可以用過水川燙法，燙過再炒。但沸水的溫度較油溫低，所以要多用些水，水滾再放入魚片。

炒魚片沒有刺，不但顯得精緻，老人家和孩子吃起來也放心，搭配上不同的配料更可以均衡營養，口味也可以變化，真是烹調魚的好選擇！

一般推薦可用於炒的魚種有

鮸魚 · 石斑切片 · 黃魚 · 養殖石斑 · 鱈魚 · 草魚中段 · 鱸魚 · 潮鯛魚片

雪菜炒魚片
Stir-fried Fish with Preserved Mustard Green

材料
潮鯛魚肉 250 公克、雪裡蕻 250 公克、
熟筍片 10 ～ 15 片、火腿絲 1 大匙（隨意）
蔥 1 支、薑片 5 ～ 6 片

調味料
Ⓐ鹽 1/4 茶匙、水 2 大匙、糖 1/4 茶匙、胡椒粉少許、蛋白 1 大匙、太白粉 1/2 大匙
Ⓑ水 3 大匙、鹽、糖各少許
Ⓒ太白粉水適量、麻油少許

做法
1. 修除紅色魚肉再斜刀片成薄片，先用鹽和水拌勻，再拌上其他調味料Ⓐ，醃 30 分鐘。
2. 雪裡蕻沖洗乾淨後切碎，擠乾水分（尾端老葉不要）。蔥切段。
3. 燒熱 2 杯油後，將魚片過油快炒至 9 分熟，撈出。
4. 用 2 大匙油爆香蔥段和薑片，放入雪裡蕻、筍片和調味料Ⓑ炒勻，加入火腿絲、魚片小心拌炒，勾下薄芡，滴下麻油即可裝盤。

Ingredients
250g. fish fillet(sea bream), 250g. preserved mustard green, 10~15 pieces cooked bamboo shoots, 1T. shredded ham(optional), 1stalk green onion, 5~6 slices ginger

Seasonings
Ⓐ1/4t. salt, 2T. water, 1/4t. sugar, a pinch of pepper, 1T. egg white, 1/2T. cornstarch
Ⓑ3T. water, a little of salt and sugar
Ⓒcornstarch paste, a little of sesame oil

Procedure
1. Slice fish, mix with salt and water first, then mix with other seasoningsⒶ, marinate for 30 minutes.
2. Rinse mustard green thoroughly, chop and squeeze out the liquid(discard rough leaves). Cut green onion into sections.
3. Heat 2C. oil, stir-fry fish until done, drain.
4. Heat 2T. oil to stir-fry green onion sections and ginger, add mustard green, bamboo shoots and seasoningsⒷ, stir evenly, put fish in, mix carefully, thicken with cornstarch paste, drop sesame oil, Remove to plate.

碧綠滑斑球

Stir-fried Fish Cantonese Style

材料

石斑魚肉 450 公克、蔥段 12 小段、薑 8 小片
胡蘿蔔花片 10 片、青花菜 1 棵

調味料

Ⓐ鹽 1/4 茶匙、胡椒粉少許、太白粉 1/2 大匙
薑汁 1 茶匙

Ⓑ酒 1/2 大匙、清湯 1/3 杯、太白粉水 2 茶匙
鹽 1/4 茶匙、麻油少許

做法

1. 將魚肉切成 3 公分大小、1.5 公分的厚片，用調味料Ⓐ醃 10 分鐘以上。
2. 青花菜分成小朵，用滾水燙熟，泡入冷水中，另起油鍋炒青花菜，加鹽和清湯煮至入味，排在盤邊。
3. 燒 2 杯油至 7 分熱後落魚球泡至 8 分熟，撈出。另用 1 大匙油爆香蔥段及薑片，落下魚球和胡蘿蔔片，淋下調味料Ⓑ，輕輕拌和即可裝盤。

Ingredients

450g. fish fillet(grouper), 12 pieces green onion section, 8 slices ginger, 10 pieces cooked carrot, 1 broccoli

Seasonings

Ⓐ1/4t. salt, a pinch of pepper, 1/2T. cornstarch, 1t. ginger juice

Ⓑ1/2T. wine, 1/3C. soup stock, 1/4t. salt, few drops of sesame oil, 2t. cornstarch paste

Procedure

1. Cut fish into 3×3cm pieces, and about 1.5cm thick, marinate with seasoningsⒶ for 10 minutes.
2. Trim broccoli, boil in boiling water for 1 minutes, drain and soak in cold water immediately. Heat 1T. oil to stir-fry broccoli, add salt and 1/2cups of soup stock, cook to the tenderness you like, drain, arrange on the serving plate.
3. Heat 2 cups of oil to fry fish slices to 80% done, drain. Heat another 1T. oil to stir-fry green onion and ginger, add fish, carrots and seasoningsⒷ, mix evenly and serve.

蠔油魚片
Fish Slices with Oyster Sauce

材料
石斑魚肉 250 公克、熟筍片、熟胡蘿蔔片、青豆、草菇各適量、蔥 2 支
薑片 8 小片

調味料
Ⓐ鹽、糖各 1/4 茶匙、薑汁 1/2 茶匙、蛋白 1 大匙、太白粉 1/2 大匙
Ⓑ蠔油 2 大匙、糖 1/4 茶匙、水 4 大匙、太白粉 1/2 茶匙、麻油少許

做法
1. 將魚肉片切成約 0.5 公分的厚片，用調味料Ⓐ拌勻醃 30 分鐘左右。
2. 草菇和青豆分別用滾水川燙一下，泡冷水備用。蔥切段。
3. 1 杯油先熱至 8 分熱，放下魚片過油至 9 分熟，撈出。
4. 用 2 大匙油爆香蔥段和薑片，放下筍片等配料炒數下，再將魚片放入鍋中，淋下調味料Ⓑ，輕輕拌炒均勻便可。

Ingredients
250g. fish fillet(grouper), cooked bamboo shoot slices, cooked carrot slices, snow peas, straw mushrooms, 2stalks green onion, 8slices ginger

Seasonings
Ⓐ1/4t. salt, 1/4t. sugar, 1/2t. ginger juice, 1T. egg white, 1/2T. cornstarch
Ⓑ2T. oyster sauce, 1/4t. sugar, 4T. water, 1/2t. cornstarch, a few drops of sesame oil

Procedure
1. Slice fish into 0.5cm thick slices, marinate with seasoningsⒶ for 30 minutes.
2. Blanch straw mushrooms and snow peas , drain and rinse with cold water. Cut green onion into sections.
3. Heat 1C. oil to 160°C, fry fish until just done, drain.
4. Heat 2T. oil to stir-fry green onion and ginger first, add vegetables and fish, pour mixed seasoningsⒷ, stir-fry carefully and quickly until evenly mixed. Remove to the serving plate.

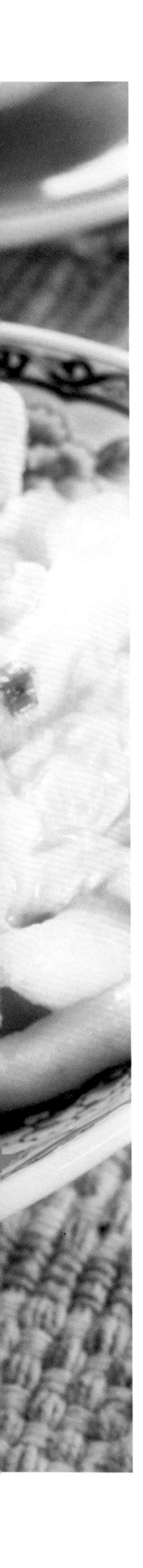

滑蛋魚絲
Stir-fried Fish with Eggs

材料
魚肉 150 公克、蛋 4 個、鹽 1/2 茶匙、蔥花 2 大匙、油 1 杯

調味料
鹽 1/4 茶匙、蛋白 1 大匙、太白粉 1 茶匙

做法
1. 將魚肉順紋切成直絲，用調勻的調味料拌勻，醃 10 分鐘以上。
2. 蛋加鹽 1/2 茶匙打散。
3. 油燒熱，將魚絲盡量分開、分散放入油中，待熟後即撈出，瀝淨油。
4. 另外將 4 大匙油燒熱，落下蔥花爆香，馬上將蛋汁淋下，用炒鏟貼住鍋底，順向滑炒，直到蛋汁大半凝固時加入魚絲，過 2～3 秒鐘便可關火，裝入碟中。

Ingredients
150g. fish fillet, 4 eggs, 1/2t. salt, 2T. chopped green onion, 1C. oil

Seasonings
1/4t. salt, 1T. egg white, 1t. cornstarch

Procedure
1. Cut fish fillet into long strips according to the grain, marinate with mixed seasonings for 10 minutes.
2. Beat the eggs well with 1/2 t. salt.
3. Heat 1C.oil, add fish strips in, make them separately as possible, fry for 10 seconds, drain.
4. Heat another 4T. oil to stir-fry green onion, when fragrant, add eggs, stir-fry in one direction over high heat, when the eggs are almost done, add fish, mix for 2~3 seconds, turn off the heat, place on plate.

魚香炒魚片

Stir-fried Fish Slices with Yu-Hsiang Sauce

材料

潮鯛魚片 250 公克、絞肉 3 大匙、木耳屑 2 大匙、荸薺 4 粒、大蒜屑 1 茶匙
薑末 1 茶匙、蔥屑 1 大匙

調味料

Ⓐ鹽 1/4 茶匙、水 2 大匙、蛋白 1 大匙、太白粉 1 大匙
Ⓑ辣豆瓣醬 1 大匙、醬油 1 大匙、糖 2 茶匙、醋 1/2 大匙、鹽 1/4 茶匙
Ⓒ水 1/2 杯、太白粉水約 1/2 大匙、麻油 1 茶匙、胡椒粉少許

做法

1. 魚肉打斜切成厚片，用調味料Ⓐ的鹽和水先拌勻，再加蛋白拌勻，至蛋白都吸收後，拌上太白粉，放入冰箱中冷藏 20 ～ 30 分鐘。
2. 煮滾水 4 杯，放入魚片，以中火燙煮 40 ～ 50 秒鐘，至魚變色已熟時，小心撈出。
3. 鍋中熱 2 大匙油，先炒香絞肉、大蒜屑和薑末，加入木耳屑、荸薺和辣豆瓣醬等調味料Ⓑ一起炒勻，加入水和魚片煮滾，勾芡後滴下麻油，撒下蔥屑和胡椒粉一拌即關火，起鍋裝盤。

Ingredients

250g. fish fillet(sea bream), 3T. ground pork, 2T. chopped black fungus, 4 water chestnut, 1t. chopped garlic, 1t. chopped ginger, 1T. chopped green onion

Seasonings

Ⓐ1/4t. salt, 2T. water, 1T. egg white, 1T. cornstarch
Ⓑ1T. hot bean paste, 1T. soy sauce, 2t. sugar, 1/2T. vinegar, 1/4t. salt
Ⓒ1/2C. water, about 1/2T. cornstarch paste, 1t. sesame oil, a pinch of pepper

Procedure

1. Cut the fish fillet into thick pieces, mix with salt and water first, then stir egg white in, when all observed , mix with cornstarch, marinate for 20~30minutes, store in refrigerator.
2. Bring 4 cups of water to a boil, blanch fish in, cook over medium heat for about 40~50 seconds, until fish turns white and done, drain.
3. Heat 2 T. oil to stir-fry ground pork, garlic and ginger, when fragrant, add fungus, water chestnut, hot bean paste and all seasoningsⒷ, stir-fry evenly, add fish and water, thicken with cornstarch paste, drop sesame oil, sprinkle green onion and pepper, mix and remove to a plate.

糟溜魚片

Fish Slices with Wine Sauce

材料
新鮮魚肉 300 公克、木耳（泡發）1 杯

調味料
Ⓐ蛋白 1 大匙、鹽 1/4 茶匙
太白粉 1 大匙
Ⓑ薑末 1/2 大匙、糖 1 大匙、清湯 1 杯
大蒜末 1 大匙、鹽 1/2 大匙
Ⓒ香糟酒 3 大匙、太白粉水 1/2 大匙

做法
1. 魚肉切成一寸多四方、半公分厚之片狀，用調味料Ⓐ抓拌，醃約 30 分鐘。
2. 油 2 杯燒至 6 分熱，倒下魚片泡至 9 分熟，撈出並瀝乾油份。
3. 木耳泡軟、摘好，用開水川燙一下，瀝乾放在盤中墊底。
4. 在炒鍋內放下調味料Ⓑ煮滾，落魚片下鍋，加入香糟酒並搖動炒鍋勾芡，使粉與汁完全混合，再淋下 1 大匙熱油，即盛裝到木耳上。

Ingredients
300g. fish fillet, 1C. soaked black fungus

Seasonings
Ⓐ1T. egg white, 1/4t. salt, 1T. cornstarch
Ⓑ1/2T. chopped ginger, 1T. chopped garlic, 1T. sugar, 1/2T. salt, 1C. soup stock
Ⓒ3T. sweet fermented rice, 1/2T. cornstarch paste

Procedure
1. Cut fish fillet into 1" pieces and about 1/4" thick. Marinate withⒶ for 30 minutes.
2. Heat 2C. oil to 120°C, add fish in, fry over medium heat, remove when it is done.
3. Trim the soaked fungus, blanch and drain, place on the serving plate.
4. Bring Ⓑ to a boil, add fish and fermented rice in, when boils again, thicken with cornstarch paste, add 1T. heated oil at last, place on top of fungus, serve.

豉椒炒魚球

Stir-fried Fish Fillet with Fermented Beans

材料
石斑魚肉 250 公克、青椒 1 小個、紅辣椒 1 支
紅甜椒 1/4 個、豆豉 1 大匙、蔥花 1 大匙
薑、蒜末各 1 茶匙

調味料
Ⓐ鹽 1/4 茶匙、水 1 大匙、蛋白 1 大匙
太白粉 1/2 大匙
Ⓑ酒 1/2 大匙、糖 1/4 茶匙、麻油 1/2 茶匙
水 3 大匙、醬油 1/2 大匙、胡椒粉少許
太白粉水 1～2 茶匙

做法
1. 魚肉切成 2×3 公分大小、1.5 公分厚的塊，用調味料Ⓐ拌勻醃好。
2. 青、紅椒及紅辣椒均去籽切片，豆豉泡水 3～5 分鐘，瀝乾。
3. 魚球先過油炒 9 分熟，撈出。油倒出，僅用約 1 大匙油炒香豆豉和薑、蒜末，加入魚球、青、紅椒、紅辣椒和調味料Ⓑ，拌炒數下均勻後，撒下蔥花，起鍋裝盤。

Ingredients
250g. fish fillet(grouper), 1green pepper, 1/4 red bell pepper, 1red chili, 1T. f fermented black beans, 1T. chopped green onion, 1t. chopped ginger, 1t. chopped garlic

Seasonings
Ⓐ1/4 t. salt, 1 T. water, 1 T. egg white, 1/2 T. cornstarch
Ⓑ1/2 T. wine, 3 T. water, 1/2 T. soy sauce, 1/4 t. sugar, 1/2 t. sesame oil, a pinch of pepper, 1~2 t. cornstarch paste

Procedure
1. Cut fish fillet into 2×3 cm, 1.5cm thick pieces. Marinate with seasonings Ⓐ.
2. Remove seeds from peppers and red chili, cut into cubes. Soak fermented black beans for 3~5 minutes, drain.
3. Heat 2.T.oil, run fish through oil until done, drain. Pour away oil, using only 1T. oil to stir-fry black beans , garlic and ginger, when fragrant, add fish, peppers, red chili and seasonings Ⓑ, stir-fry until evenly mixed. Sprinkle green onion, transfer to plate.

鮮茄燴炒魚片

Fish Slices with Tomato

材料

石斑魚肉 250 公克、蕃茄 2 個、木耳少許、菠菜 150 公克、蔥 2 支

調味料

Ⓐ鹽少許、水 2 大匙、太白粉 1 大匙、蛋白 1 大匙
Ⓑ醬油 1/2 大匙、鹽 1/3 茶匙、糖 1/2 茶匙、水 3/4 杯、太白粉水適量

做法

1. 魚肉切除紅色部分，再打斜切片，用調味料Ⓐ中的鹽和水先抓拌一下，再拌入蛋白和太白粉，拌勻醃 30 分鐘。
2. 蕃茄切塊，木耳撕成小片，菠菜切成 1 吋長段，蔥切段。
3. 魚肉先過油或用熱水川燙至 9 分熟，撈出。
4. 用 2 大匙油爆香蔥段，放下蕃茄翻炒至微軟，加入調味料Ⓑ和木耳煮片刻，放下菠菜炒軟後盛放盤中。把魚片放入湯汁中一滾即勾芡，盛到蔬菜料上。

Ingredients

250 g. fish fillet(grouper), 2 tomatoes, soaked black fungus, 150 g. spinach, 2 stalks green onion

Seasonings

Ⓐa little of salt, 2 T.water ,1 T. cornstarch, 1 T. egg white
Ⓑ1/2 T. soy sauce, 1/3 t. salt, 1/2 t. sugar, 3/4 C. water, cornstarch paste

Procedure

1. Discard the red parts of fish,then slice into thin pieces. Mix with salt and water first, then add egg white and cornstarch.Marinate for 30 minutes.
2. Cut tomatoes into pieces, trim black fungus, cut spinach into 1 inch long, cut green onion into sections.
3. Run the fish through heated oil or boiling water, drain.
4. Heat 2 T. oil to stir-fry green onion sections, when fragrant, add tomatoes, stir-fry again until tomatoes become soft. Add Ⓑ and fungus, cook for a while. Add spinach in. Remove them to a plate when spinach gets soft. Add fish to sauce, thicken with cornstarch paste when it boils. Place on top of the vegetables.

銀芽韭黃炒鱔魚
Stir-fried Eel with Bean Sprouts and Leek

材料
鱔魚 250 公克、銀芽 150 公克
韭黃 100 公克、嫩薑絲 2 大匙

調味料
Ⓐ醬油 1 大匙、胡椒粉少許、太白粉 1 大匙
Ⓑ酒 1 大匙、糖 1 茶匙、太白粉水適量
深色醬油 1 大匙、鎮江醋 1/2 大匙
胡椒粉 1/6 茶匙

做法
1. 鱔魚切成兩長段後，用滾水燙 5 秒鐘即撈出，沖涼一些，切成 4 公分的粗絲（如筷子般粗細），用調味料Ⓐ拌勻，醃 20 分鐘。（一般魚肉只要切粗條醃入味即可）
2. 銀芽用滾水川燙去生味，撈出。韭黃切段。
3. 用 9 分熱的油將鱔魚過油，撈出。
4. 用 1 大匙油先炒薑絲，隨即加入鱔魚、銀芽和調味料Ⓑ，大火炒均勻，關火後拌入韭黃，拌勻裝盤。

Ingredients
250 g. small eels, 150 g. bean sprouts, 100 g. yellow leek, 2 T. shredded ginger.

Seasonings
Ⓐ1 T. soy sauce, a pinch of pepper, 1 T. cornstarch
Ⓑ1 T. dark soy sauce, 1 T. wine, 1 t. sugar, 1/2 T. brown vinegar, 1/6 t. pepper, cornstarch paste

Procedure
1. Cut eel half, blench for 5 seconds, drain. Rinse with cold water, cut into 4 cm long strings, marinate with Ⓐ for 20 minutes.
2. Blanch bean sprouts, drain. Cut yellow leek into sections.
3. Heat oil to 180°C, run eel through oil, drain.
4. Heat 1 T. oil to stir-fry ginger shreds, add eel, bean sprouts and Ⓑ. Stir-fry over high heat until evenly mixed, turn off the heat, add yellow leek in, mix again. Serve hot.

辣炒丁香
Spicy Dried Fish

材料
丁香魚 120 公克、豆腐乾 8 片、蔥 2 支
大蒜屑 1 大匙、青蒜 1 支、紅辣椒 2 支
綠辣椒 2 支

調味料
醬油 1 大匙、鹽適量、糖 2 茶匙
水 4 大匙

做法
1. 丁香魚洗淨，晾乾水分，用 4 大匙油慢慢煸炒至酥透，盛出。
2. 豆腐乾切粗條，蔥、青蒜、辣椒等辛香料切成斜段。
3. 先將豆腐乾用 3 大匙油煎黃外層，加入大蒜屑等辛香料爆香，加調味料炒勻，放回丁香魚翻炒數下，至汁收乾即可關火。

Ingredients
120g. small dried fish, 8 pieces dried bean curd, 2 red chili, 2 green chili, 1T. chopped garlic, 1 green garlic, 2 stalks green onion

Seasonings
1T. soy sauce, salt, 2t. sugar, 4T. water

Procedure
1. Rinse the fish, pat dry. Stir-fry with 4T. oil gradually until the fish becomes crispy, remove.
2. Cut dried bean curd into strips, cut other ingredients into sections.
3. Stir-fried dried bean curd with 3T. oil until outside becomes brown, add other ingredients and seasonings, stir-fry evenly, add fish, stir-fry again. Turn off the heat when the liquid absorbed.

Stew 燒

在「燒」之前——

「燒」是家庭中常用來烹調魚的方法，細探起來，也是變化多多的。辛香料和配料用的不同，就會改變燒出來的味道。

通常在紅燒魚的時候總是先煎一下，增加香氣後再燒。因此在煎之前，最好也要用油盪一下鍋，以免皮破不好看，一翻身肉也碎了，還沒燒就不成型了。魚煎過之後，要把辛香料蔥薑等爆香，再淋酒和醬油，等燒個幾秒鐘後再加水，使酒和醬油的香氣能遇熱發揮出來。

加水的量和燒的時間有關，也就是和魚的大小有關，同時也和個人喜好有關。有的人喜歡魚燒得嫩嫩的，就可以把汁勾個芡，使味道能附著在魚身上。有的喜歡燒久一點使魚入味。常聽老一輩的說：「千燉豆腐，萬燉魚」，表示魚要燉很久才入味好吃。在魚身上劃切刀口，就是幫助入味的方法之一；煎得透一點再燒，也可以快些吸收滋味，當然大的魚頭剁開來燒，一定比整個燒來得快。燒魚是要有些耐性和時間才燒得好吃的。

餐廳中燒魚，為了要搶時間，常常是用蒸的，起鍋後再淋上炒過的味汁。蒸的魚不會破皮，賣相好，又不用佔個鍋子可以省空間。但是燒魚講究要「入味」的效果就差一點了。

燒魚，總要搭配辛香料來幫助去腥、添味的，例如蔥、薑、蒜、辣椒都是常用的，另外一些醃漬的材料，如雪裡蕻、梅乾菜、豆豉、豆瓣醬都有特殊風味。在使用辛香料時，要先把它們在油中爆香，以增加更好的效果。「爆香」，看似簡單，但也要注意火候；辛香料中含有水分，要先大一點火除去水分，再小一點火煎焦黃一點，產生香氣。如此辛香料才能發揮最好的效果。

幾乎任何魚都可以紅燒，至於什麼魚要搭配什麼口味來燒，也看個人喜歡。其實不同魚肉的口感、味道，就會改變燒出來的滋味，例如蔥燒鱈魚和蔥燒鱸魚，調味料和辛香料一樣，味道卻是不一樣的。基本燒魚的重點抓住了，魚是怎麼燒怎麼好吃的。

一般推薦可用於燒的魚種有

草魚尾 · 馬加魚切片 · 鮭魚頭 · 午仔頭 · 青衣 · 豆仔魚 · 鯧魚 · 帶魚
赤鯮魚

大蒜燒黃魚
Stewed Fish with Garlic

材料
大黃魚 1 條（約 500 公克重）
香菇 3 朵、粉皮 2 張、薑片 2 片
蔥 10 小段、大蒜粒 10 粒、青蒜 1/2 支

調味料
酒 1 大匙、醬油 4 大匙、糖 1 大匙
水 3 杯、胡椒粉少許

做法
1. 黃魚兩面各切兩、三條斜刀痕（也可以切成 2 段）。香菇泡軟切片。青蒜切絲。
2. 用熱油把黃魚兩面煎黃，盛出魚。放入大蒜爆香，再放入蔥段、薑片煎香，加入香菇和調味料，先以大火煮滾，再改小火，蓋好鍋蓋，燒約 20 分鐘至湯汁剩下 1 杯左右。
3. 加入切寬條的粉皮，煮至粉皮透明，撒下切好之青蒜絲即可裝盤。

Ingredients
1 fresh fish(about 500g.), 3 black mushroom, 2 pieces mung bean sheet, 2 slices ginger, 10 green onion sections,10 cloves garlic, 1/2 green garlic

Seasonings
1T. wine, 4T. soy sauce, 1T. sugar, 3C. water, a pinch of pepper

Procedure
1. Score 2~3 times on both side of fish or cut the fish into two sections. Soak and slice black mushrooms. shred green garlic.
2. Heat 5T. oil to fry both sides of the fish, Remove. Add garlic, fry until brown and fragrant, add green onion and ginger, stir-fry together, add black mushroom and seasonings, bring to a boil over high heat, reduce to low, simmer for 20 minutes until the liquid reduce to 2/3 cup.
3. Add mung bean sheet sections, cook until it becomes soft, sprinkle shredded green garlic in, remove to the serving plate.

蔥燒鮮魚
Stewed Fish with Green Onion

材料
新鮮魚 1 條、約 12 兩、蔥 3 支
薑絲 1 大匙

調味料
鹽 1/4 茶匙、酒 2 大匙、醬油 2 大匙
糖 2 茶匙、烏醋 1/2 大匙、水 2 杯

做法
1. 魚身兩面切 2 ～ 3 條刀口，如選用比較扁平、肉薄的魚，也可以不切刀口。蔥切段。
2. 鍋中燒熱油 3 大匙，放入魚以中火煎黃表面，翻面再煎時，加入蔥段和薑絲一起煎香。
3. 蔥段夠焦黃時，淋下調勻的調味料，煮滾後改小火煮約 15 ～ 20 分鐘（中途翻面一次），至湯汁約剩 1/2 杯時即可關火。

Ingredients
1 fish ,about 450g, 3 stalks green onion, 1T. shredded ginger

Seasonings
1/4t. salt, 2T. wine, 2T. soy sauce, 2t . sugar, 1/2T. vinegar, 2C. water

Procedure
1. Score 2~3 cuts on both sides if needed, if the fish you cook is thin and flat, no need to scored. Cut green onion into sections.
2. Heat 3T. oil, fry the fish until both sides get brown, add green onion and ginger to fry together when you fry the second side.
3. When the green onion becomes brown, add seasonings, bring to a boil, simmer for 15~20minutes(turn over the fish once while simmer). Turn off the heat when the liquid reduce to half cup.

五味豆仔魚
Stewed Fish with Spices

材料
豆仔魚 3 條、蔥花 2 大匙、薑末 1/2 大匙
大蒜末 1 大匙、紅椒末 1/2 大匙
香菜段少許

調味料
Ⓐ酒 1 大匙、醬油 2 大匙、糖 1 大匙
鹽 1/4 茶匙、水 1 杯
Ⓑ烏醋 1 大匙、麻油數滴

做法
1. 魚洗淨擦乾，抹少許酒，入鍋中用 3 大匙油煎至半熟。
2. 加入薑末和蒜末一起爆香，淋下調味料Ⓐ燒約 5 ～ 6 分鐘。
3. 待魚熟後，撒下紅椒末、蔥花和調味料Ⓑ烹香，起鍋時再撒下香菜段。

Ingredients
3 broneo mullet , 2T. chopped green onion, 1/2T. chopped ginger, 1T. chopped garlic, 1/2T. chopped red chili, cilantro sections

Seasonings
Ⓐ 1T. wine, 2T. soy sauce, 1T. sugar,1/4t.salt, 1C. water
Ⓑ 1T. black vinegar, a few drops of sesame oil

Procedure
1. Rinse fish, pat dry. Brush wine all over fish, fry with 3T. oil until half done.
2. Add chopped ginger and garlic in, pour seasonings Ⓐ in, cook for 5~6 minutes.
3. When fish is done, sprinkle red chili, green onion and seasonings Ⓑ, add cilantro just before turn off the heat.

雪菜燒帶魚
Stewed Fish with Mustard Green

材料
帶魚 1 條，約 12 兩、雪裡蕻 4 兩
蔥 2 支、薑 2 片、紅椒 1 支

調味料
酒 1 大匙、醬油 2 大匙、糖 1 大匙
水 1½ 杯

做法
1. 帶魚切成約 5 公分長的塊，每塊上切數條刀口。蔥切段；紅椒切圈。
2. 雪裡蕻沖洗乾淨，以免有沙，切碎後擠乾水分。
3. 用 2 大匙油煎黃帶魚的兩面，魚盛出。另外加 1 大匙油爆香蔥段和薑片，放入雪裡蕻和調味料，煮滾後放入帶魚，改小火燒約 15 ～ 20 分鐘，撒下紅椒圈。若湯汁仍多，以大火收乾一些即可。

Ingredients
1 hair tail,about 450g, 150g. preserved mustard green, 2 stalks green onion, 2 slices ginger, 1 red chili

Seasonings
1T. wine, 2T. soy sauce, 1T. sugar, 1½C. water

Procedure
1. Cut the hair tail into 5 cm pieces, score a few times one each side. Cut green onion into sections. Cut red chilli into rings.
2. Rinse mustard green thoroughly, chop and squeeze out the liquid.
3. Heat 2T. oil to fry both sides of the fish, remove. Add another 1T. oil to fry green onion and ginger. Add mustard green and seasonings, put the fish in after boiled. Simmer for 15~20 minutes. Add red chili slices. Turn to high heat if the liquid is too much.

梅菜燒魚
Stewed Fish with Fermented Cabbage

材料
石斑魚 1 條（約 500 公克）、梅乾菜 50 公克、蔥 2 支、香菜少許

蒸梅菜料
醬油 2 大匙、酒 1 大匙、冰糖 1 大匙、八角 1/2 顆，水 1 杯、太白粉水適量

做法
1. 梅乾菜泡水洗淨細沙，擠乾水分後切成 1 公分碎段。放碗中加蒸料蒸 1 小時，至菜料已軟。
2. 魚打理乾淨，兩面均切劃刀口。蔥切段。
3. 起油鍋用 2 大匙油將魚和蔥段煎香，梅乾菜連汁一起倒入鍋中，可酌量加水同煮 15 分鐘，將汁收至 2/3 杯左右，湯汁用少許太白粉水略勾芡，關火、撒下香菜，裝盤。

Ingredients
1 grouper, 500g. fermented cabbage, 2 stalks green onion, a little of cilantro

Seasonings
2T. soy sauce, 1T. wine, 1T. rock sugar, 1/2 star anise, 1C. water, cornstarch paste.

Procedure
1. Soak fermented cabbage in water, rinse several times to clean all sands, squeeze out the water, chop into small pieces. Place in a bowl, steam with seasonings for 1 hour till cabbage turn soft.
2. Rinse fish, score a few cuts on both sides. Cut green onion into sections.
3. Heat 2T. oil to fry fish and green onion, add cabbage and the juice in(you may add some water if need), stew for 15 minutes until juice reduce to 2/3cup. Thicken with a little cornstarch paste. Turn off the heat, sprinkle cilantro sections over, remove to a plate.

五柳枝
Fish with Vegetable Sauce

材料
午仔魚 1 條、薑絲 1 大匙、肉絲、白菜絲
香菇絲、胡蘿蔔絲、筍絲各酌量、蔥 1 支

調味料
Ⓐ鹽 1/4 茶匙、淡色醬油 1 大匙、糖 2 大匙
醋 3 大匙、水 1 杯
Ⓑ麻油1/2茶匙、胡椒粉1/6茶匙、太白粉水酌量

做法
1. 午仔魚清理乾淨，在魚身上劃切刀痕，撒下少許鹽醃約 5 分鐘。用約 3 大匙油煎黃魚的兩面，盛出。
2. 肉絲用少許醬油、水和太白粉抓拌醃一下。蔥切絲。
3. 另用 2 大匙油炒肉絲和香菇絲，再放入白菜絲同炒，待白菜微軟，加入魚、胡蘿蔔絲、筍絲和調味料，煮滾後，改小火燒 7～8 分鐘，先盛出魚，菜料中放蔥、薑絲後用少許太白粉水勾芡，淋下麻油及胡椒粉，全部淋到魚上即可。

Ingredients
1 thread fin, pork strings, Chinese cabbage, black mushroom, carrot, bamboo shoot, 1stalk green onion, 1T. shredded ginger, 1 red chili

Seasonings
Ⓐ1/4t. salt, 1T. light color soy sauce, 2T. sugar, 3T. vinegar, 1C. water
Ⓑ1t. sesame oil, 1/6t. pepper, cornstarch paste

Procedure
1. Rinse fish, score a few times on fish, sprinkle a little of salt on fish, marinate for 5 minutes. Fry the fish with 3T. oil until both sides get brown, remove.
2. Marinate pork strings with soy sauce, water and cornstarch for a while. Shred green onion .
3. Stir-fry pork and black mushroom with 2T. oil, add cabbage in, stir-fry until cabbage becomes soft. Add fish, carrot, bamboo shoot and seasonings, after it boils, simmer for 7~8 minutes. Remove fish first, add green onion and ginger in, thicken sauce with a little cornstarch paste. Drop sesame oil and pepper, pour the sauce over fish, serve.

辣豆瓣魚
Carp with Hot Bean Sauce

材料
活鯉魚 1 條或其他活魚亦可、薑屑 1 大匙
豆腐 1 塊、大蒜屑 1 大匙、酒釀 1 大匙

調味料
Ⓐ辣豆瓣醬 2 大匙、酒 1 大匙、醬油 2 大匙
鹽 1/2 茶匙、糖 2 茶匙、水 2 杯
Ⓑ太白粉水少許、麻油 1 茶匙、蔥花 2 大匙
鎮江醋 1/2 大匙

做法
1. 鯉魚打理乾淨後，擦乾水分，在魚身上斜切 2～3 條刀紋。
2. 鍋中燒熱油 4 大匙，將魚的兩面稍微煎一下，盛出。放入薑、蒜屑爆香，再放入辣豆瓣醬和酒釀同炒，淋下調味料Ⓐ一起煮滾，放回魚和豆腐，一起燒煮約 10 分鐘。
3. 見汁已剩一半時，將魚和豆腐盛出裝盤。湯汁勾芡，並加入調味料Ⓑ炒匀，把汁淋在魚身上。

Ingredients
1 carp or other live fish, 1piece bean curd, 1T. chopped ginger, 1T. chopped garlic, 1T. sweet fermented rice

Seasonings
Ⓐ2T. hot bean paste, 1T. wine, 2T. soy sauce, 1/2t. salt, 2t. sugar, 2C. water
Ⓑa little of cornstarch paste, 1/2T. brown vinegar, 1t. sesame oil, 2T. green onion

Procedure
1. Rinse and pat dry the carp. Score 2~3 times on both sides.
2. Heat 4T. oil in wok, fry the fish lightly, remove. Add ginger and garlic, fry until fragrant, add hot bean paste ,fermented rice, and all seasoningsⒶ, bring to a boil, return carp and bean curd in, stew for 10 minutes.
3. When juice reduce to half, remove carp and bean curd to a plate. Thicken the juice, addⒷ, mix evenly, pour over carp.

紅燒划水
Stewed Fish Tail Shanghai Style

材料
草魚尾 1 段、蔥 2 支、薑 2 片、太白粉 1 大匙
青蒜絲 1/2 杯

調味料
Ⓐ醬油 2 大匙、胡椒粉 1/6 茶匙
Ⓑ酒 1/2 大匙、深色醬油 3 大匙、糖 1 大匙
水 1$\frac{1}{2}$ 杯

做法
1. 草魚尾打理乾淨，直剖成 5 條，用調味料Ⓐ拌醃 5 分鐘。蔥切段。
2. 太白粉 1 大匙加水 3 大匙調稀。
3. 炒鍋中燒熱 3 大匙油，放入蔥段、薑片，煎黃後即取出丟棄。魚尾沾太白粉水，放入油中輕輕煎過，排在鍋中，淋下調味料Ⓑ，蓋上鍋蓋，用中小火燒約 5 分鐘，至魚肉已熟透。
4. 輕輕勾芡後，淋下 1 大匙熱油，撒下青蒜絲，全部滑入大盤中，可再撒下少許胡椒粉上桌。

Ingredients
1 grass carp tail, 2 stalks green onion, 2 slices ginger, 1T. corn starch, 1/2C. green garlic sheds

Seasonings
Ⓐ2T. soy sauce, 1/6t. pepper
Ⓑ1/2T. wine, 3T. dark color soy sauce, 1T. sugar, 1½C. water

Procedure
1. Rinse fish, cut into 5 long strips. Mix with seasoningsⒶ for 5 minutes. Cut green onion into sections.
2. Dissolve 1T. cornstarch with 3T. water.
3. Heat 3T. oil to fry green onion and ginger, remove when it gets brown. Dip fish tail with cornstarch past, fry slightly then arrange in wok, add seasoningsⒷ, cover. Cook over medium low heat for 5 minutes until fish is done.
4. Thicken with cornstarch paste, add 1T. heated oil, sprinkle green garlic shreds, remove to plate, you may add a little of pepper on top.

紅燒青衣
Stewed Fish with Soy Sauce

材料
青衣 1 條（約 500 公克）、蔥 3 支
薑 3 片

調味料
酒 2 大匙、醬油 3 大匙、糖 1/2 大匙
醋 1 茶匙、水 2 杯、麻油數滴

做法
1. 魚打理乾淨、劃上刀口。蔥切段。
2. 燒熱 3 大匙油，將魚的兩面略煎過，將魚推到鍋邊，放入蔥段和薑片煎香，把魚推回鍋中，淋下酒和醬油煮一滾，再加入糖、醋和水，煮滾後改小火慢煮。
3. 約 10 分鐘後翻面再燒，燒的時候將湯汁不斷地淋到魚身上，再燒約 6～7 分鐘即可滴下麻油裝盤。

Ingredients
1 green wrasse(about 500g.), 3 stalks, 3slices ginger

Seasonings
2T. wine, 3T. soy sauce, 1/2T. sugar, 1t. vinegar, 2C. water, a few drops of sesame oil

Procedure
1. Rinse fish. Score 2~3 times on each side of fish. Cut green onion into sections.
2. Heat 3T. oil, fry fish slightly, push the fish aside, add green onion and ginger, when it gets brown, remove the fish back. Sprinkle wine and soy sauce, cook for a while. Add sugar, vinegar and water, bring to a boil, reduce the heat.
3. Simmer for 10 minutes, turn the fish over continue to cook for 6~7minutes, pour the sauce over fish while cooking. Drop sesame oil and turn off the heat.

紅燒白果鰻
Stewed Eel with Ginkgo Nuts

材料
河鰻 1 條、白果 30 粒、紅棗 20 粒、蔥 2 支
薑 2 ～ 3 片、青蒜絲 1/2 杯

調味料
Ⓐ酒 1 大匙、醬油 3 大匙、糖 1 大匙、水 3 杯
胡椒粉少許
Ⓑ醋 1 茶匙、麻油 1/2 茶匙、太白粉水少許

做法
1. 鰻魚在 8 分熱的水中川燙至表皮變白，撈出，刷洗乾淨，切成 3 公分長段。
2. 白果沖洗一下、紅棗泡漲、蔥切段。
3. 用 2 大匙油爆香蔥段和薑片，加入調味料Ⓐ煮滾後，放入鰻魚和紅棗，以中小火煮 20 分鐘。
4. 最後 5 分鐘加入白果一同燒煮，最後淋下醋和麻油並勾芡，撒下青蒜絲即可。

Ingredients
1 eel, 30 ginkgo nuts, 20 red dates, 2 stalks green onion, 2~3 slices ginger, 1/2 C. green garlic shreds

Seasonings
Ⓐ1T. wine, 3T. soy sauce, 1T. sugar, a pinch of pepper, 3C water
Ⓑ1t. vinegar, 1/2t. sesame oil, cornstarch paste

Procedure
1. Scale the eel in 160°C hot water for 3 seconds. Remove and brush eel skin until clean. Cut into 3 cm sections.
2. Rinse ginkgo nuts. Soak red dates to soft. Cut green onion into sections.
3. Heat 2T. oil to fry green onion and ginger, add seasonings Ⓐ, bring to a boil, add eel and red dates. Cook over medium low heat for 20 minutes.
4. Add ginkgo nuts to cook together at the last 5 minutes. Drop vinegar and sesame oil. Thicken with cornstarch paste, sprinkle green garlic shreds. Serve.

蔭豉燒下巴

Stewed Fish Jaws with Black Bean Sauce

材料
潮鯛下巴 4 個、黑豆豉 1 大匙、蔥 2 支
紅辣椒 1 支、薑絲 1 大匙

調味料
Ⓐ太白粉 1 大匙、水 3 大匙
Ⓑ酒 1 大匙、醬油 1 大匙、糖 1/2 大匙

做法
1. 潮鯛下巴清洗乾淨，擦乾水分。蔥切段、辣椒切片。
2. 鍋中燒熱油 2 大匙，放下沾過太白粉水的魚頭，略煎一下（兩面均要煎過）。放下豆豉、蔥段和薑絲，淋下調味料Ⓑ和水 1 杯，煮滾後改小火煮至湯汁快收乾即可。
3. 關火後撒下辣椒片，即可裝盤。

Ingredients
4 fish jaws, 1T. fermented black beans, 2 stalks green onion, 1 red chili, 1T. ginger shreds

Seasonings
Ⓐ1T. cornstarch, 3T. water
Ⓑ1T. wine, 1T. soy sauce, 1/2T. sugar

Procedure
1. Rinse the jaws. Pat dry. Cut green onion and red chili into sections.
2. Heat 2T. oil, dip the jaws with cornstarch water, fry them with oil, add fermented black beans, green onion sections and ginger shreds, pour seasonings Ⓑ and 1C. of water. Bring to a boil, simmer until the juice reduced.
3. Add red chili slices after turn off the heat. Remove to plate.

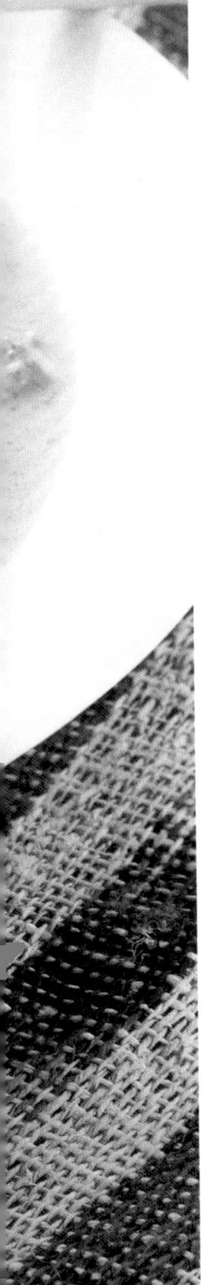

南洋咖哩魚
Fish with Curry Sauce

材料
魚肉 200 公克、芹菜 2 支、洋蔥 1/4 顆、大紅辣椒 1 條、韭菜 2 支、蛋 3 個
奶油 1 大匙、奶水 2 大匙、紅油 1 茶匙

調味料
Ⓐ鹽少許、太白粉少許
Ⓑ高湯 1 杯、蠔油 1 大匙、味露 1 茶匙、糖 1 茶匙、咖哩粉 2 茶匙

做法
1. 魚肉切片後用調味料Ⓐ拌醃 10 分鐘。
2. 芹菜、韭菜分別切段，洋蔥切絲，辣椒切斜段。
3. 紅油、奶水和蛋打均勻。
4. 炒鍋中用油將魚片先過油至 6 分熟，撈出。油倒出，僅留約 1 大匙油，將芹菜、洋蔥絲、大紅辣椒片、韭菜段和奶油炒香。
5. 加入調味料Ⓑ煮滾，放入魚片，小火煮一下，沿鍋邊淋下蛋汁，見蛋汁已熟即可盛出。

Ingredients
200g. fish fillet, 2 stalks celery, 1/4 onion, 1 red chili, 2 leek, 1T. butter, 3 eggs, 2T. milk, 1t. red chili oil

Seasonings
Ⓐa pinch of salt, a little cornstarch
Ⓑ1C. soup, 1T. oyster sauce, 1t. fish sauce, 1t. sugar, 2t. curry powder

Procedure
1. Slice fish fillet, marinate with seasoningsⒶ for 10 minutes.
2. Cut celery and leek into sections. Shred onion, slice red chili.
3. Mix red chili, milk and eggs well.
4. Heat 1C. oil, run fish slices through oil till almost done, drain. Pour oil away, using only 1T. oil to stir-fry celery, onion, red chili, leek and butter.
5. When fragrant, add seasoningsⒷ and fish, cook over low heat for a while. Pour egg mixture around the ingredients, mix and serve.

乾燒魚頭
Stewed Fish Head with Meat Sauce

材料
鰱魚頭 1/2 個、絞肉 3 大匙、蔥屑 3 大匙
薑末 1 大匙、大蒜末 1 大匙、青蒜丁適量
麻油少許

調味料
辣豆瓣醬 1/2 大匙、酒 1 大匙、醬油 3 大匙
糖 1 大匙、甜酒釀 1 大匙、胡椒粉 1/4 茶匙
甜麵醬 1/2 大匙、水 2 ½ 杯

做法
1. 魚頭先洗淨擦乾水分，浸在 2 大匙醬油中泡 10 分鐘，用熱油煎黃兩面，盛出。
2. 另用 3 大匙油炒香絞肉、薑末、蒜末和蔥屑，加入辣豆瓣醬和甜麵醬炒透，再加其他調味料，大火煮滾後放下魚頭，用小火燉煮。
3. 約 20 分鐘後翻面再燉燒，燒時並用鏟子將湯汁往魚頭上淋，至湯汁將收乾時，淋下麻油，撒下青蒜丁即可起鍋。

Ingredients
1/2 carp head, 3T. grounded pork, 3T. chopped green onion, 1T. chopped ginger, 1T. chopped garlic, diced green garlic, a few drops of sesame oil

Seasonings
1/2T. hot bean paste, 1/2T. soy bean paste, 1T. wine, 3T. soy sauce, 1T. sugar, 1T. fermented sweet rice, 1/4t. pepper, 2 ½C. water

Procedure
1. Rinse fish head, marinate with 2T. soy sauce for 10 minutes. Fry with heated oil until both sides becomes brown.
2. Heat another 3T. oil to stir-fry pork, ginger, garlic and green onion, add hot bean paste and soy bean paste, stir-fry until fragrant, add other seasonings, add fish head after the sauce boiled, simmer for 20 minutes.
3. Turn the head over, continue to cook, pour sauce over the head while cooking. When sauce is almost absorbed, drop sesame oil and sprinkle diced green garlic over. Remove and serve.

蒜燒鮭魚頭
Stewed Salmon Head with Garlic

材料
鮭魚頭 1/2 個、絞肉 2 大匙、大蒜 10 粒
紅辣椒 1 支、青蒜 1 支

調味料
酒 1 大匙、醬油 2 大匙、黑胡椒粉 1/4 茶匙
糖 2 茶匙、鹽 1/4 茶匙、醋 1/2 大匙
水 1 ½ 杯、太白粉水適量

做法
1. 鮭魚頭剁成大塊。紅辣椒和青蒜切斜段。大蒜切厚片。
2. 鍋中熱 4 大匙油，放下魚頭煎黃表面，盛出。再放入大蒜粒煎黃，同時放入絞肉同炒，淋酒並加入所有調味料（除太白粉水外）。放下魚頭，煮滾後改成小火，燒約 15 分鐘。
3. 撒下紅辣椒段和青蒜段，並以太白粉水勾芡，即可裝盤。

Ingredients
1/2 salmon head, 2T. ground pork, 10 cloves garlic, 1 red chili, 1 green garlic

Seasonings
1T. wine, 2T. soy sauce, 2t. sugar, 1/4t. salt, 1/2T. vinegar, 1/4t. black pepper, 1 ½C. water, cornstarch paste

Procedure
1. Cut salmon head into big pieces. Slice garlic, red chili and green garlic.
2. Heat 4T. oil, fry salmon until brown, remove. Add garlic and ground pork, stir-fry until fragrant. Sprinkle wine, add all seasonings(except cornstarch). Put salmon back, bring to a boil, simmer for 15 minutes.
3. Sprinkle red chili and green garlic on top, thicken with cornstarch paste, serve.

Quick boil 燙

在「燙」之前——

減肥食譜中最常用的烹調法就是「燙」。就烹調法本身來講，它是以水為媒介至熟，完全不含油，沒有熱量，是一種健康的烹調方式。當然為了使魚肉滑嫩、有好口感，處理魚片時還是要用澱粉先漿過，但是整體的熱量是少多了，希望控制體重，或有心血管問題的人，可以優先考慮用燙的烹調法。

因為是以「水」來燙熟魚片，而水的沸點在 100℃，已經比一般過油時的油溫低，因此用水燙時，水要大滾才下鍋。同時為避免由冰箱取出的魚片下鍋，會很快把水溫降低，因此水最好多一些。同時要燙的魚不怕會黏在一起，又因為沒有油的滋潤，嫩滑度會差一點，因此醃的時候用的太白粉，可以比過油的魚片多一點，當然太多也不好吃。

魚下鍋之後的火候，基本上不能太大，以免水很快再大滾，使魚肉老了。但是每一個家庭的爐火不同，有經驗的煮夫、主婦們也知道，遇到煮晚餐時的火力又特別小，因此食譜中我說「中小火」，希望大家自己拿捏。魚條、魚柳等體積小，很快就熟了，就沒有關係，越是厚片、大塊的魚肉（甚至整條魚），希望外層滑嫩、裡面熟透，就需要用小一點的火侯把魚泡熟。杭州名菜「西湖醋魚」，就是很好的例子，不論是用中段草魚片開或打斜切片，都是小火泡至熟。

燙，基本上可以做為一個前處理，燙好的魚片可以拌、可以炒、可以再加料燴、煮、烹，除了拌是直接吃，是要燙熟之外，其餘的要再烹調，因此要預留一些再熟成的空間（例如沙茶拌魚條），以免再下鍋魚就老了。

把握住「燙魚」的火候，要拌什麼味道就可以變化了。本書中其他做法裡的一些口味也可以應用，例如「炒」裡面的蠔油、豉椒、魚香、雪菜，「蒸」的 XO 醬、豆酥、蒜蓉、酸辣屑子，都適合用來拌。

要好吃又不胖，「燙」是最好的選擇！

一般推薦可用於燙的魚種有

養殖石斑 · 草魚中段 · 鯧魚 · 潮鯛魚片 · 石斑切片 · 鱸魚 · 金線魚

西湖醋溜魚

Fish with Sweet and Sour Sauce

材料

魚肉 250 公克、銀芽 150 公克、嫩薑 1 塊（切成絲約 1/2 杯；另切薑片 3 片）
蔥 1 支

調味料

Ⓐ鹽 1/4 茶匙、水 3 大匙、太白粉 1 大匙

Ⓑ糖 3 大匙、鎮江醋 4 大匙、深色醬油 2 大匙、鹽 1/4 茶匙、麻油 1/2 茶匙
水 1½ 杯、太白粉 1/2 大匙

做法

1. 去除紅色魚肉，再切成約 0.3 公分的片，用鹽和水先抓拌至產生黏性，再拌上太白粉，冷藏 30 分鐘。
2. 銀芽用滾水燙至脫生，撈出，瀝乾水分，放在盤中。
3. 另煮滾開水 4 杯（加入蔥支、薑片同煮），投入魚片，以中小火泡煮，約 40 秒左右，見魚肉已熟，撈出放在銀芽上。
4. 嫩薑絲用冰開水泡一下後，擠乾，撒在魚片上。
5. 燒熱油 2 大匙，淋下調味料Ⓑ煮滾，淋在魚上，趁熱上桌。

Ingredients

250g. fish fillet, 150g. bean sprouts, 1puece ginger(shred ginger 1/2C,cut ginger into 3 slices), a stalk green onion

Seasonings

Ⓐ1/4t. salt, 3T. water, 1T. cornstarch

Ⓑ3T. sugar, 4T. Vinegar, 2T. dark color soy sauce, 1/4. salt, 1/2t.sesame oil, 1½C. water, 1/2T. cornstarch

Procedure

1. Slice fish fillet into 0.3cm thin slices. Mix with seasoningsⒶ, store in refrigerator for 30 minutes.
2. Blanch bean sprouts, drain and place on the serving plate.
3. Boil 4C. water with green onion and ginger, add fish in, cook over medium low heat for 40 seconds, drain, place on top of bean sprout.
4. Shred ginger, soak with ice water for a while, drain and place on top of fish.
5. Heat 2T. oil, bring seasonings Ⓑ to a boil, pour the sauce over fish. Serve.

麻辣魚柳

Fish Strips with Spicy Sauce

材料

魚肉 250 公克、黃瓜 1 支、茭白筍 1 ～ 2 支、蔥絲 1/2 杯、香菜段少許
蔥、薑、酒（燙魚用）

調味料

Ⓐ鹽少許、胡椒粉少許、水 3 大匙、太白粉 1 大匙
Ⓑ麻辣汁：芝麻醬 1 茶匙、醬油 2 大匙、麻油 1/2 大匙、花椒粉 1/2 茶匙
糖 1/2 大匙、 水 2 大匙、醋 1 茶匙、紅油 1/2 大匙、蒜泥 1 茶匙

做法

1. 魚肉切成粗條，用鹽和水輕輕先抓拌一下，再拌上胡椒粉和太白粉，放約 20 分鐘。
2. 黃瓜洗淨，擦乾水分，切成薄片舖放在盤子上。茭白筍切絲。
3. 鍋中水煮滾，放下茭白筍燙一下，瀝乾水分，放在黃瓜上。水中再加蔥、薑和酒少許，放入魚柳燙熟，撈出後拌上蔥絲，放在茭白筍上，撒下香菜段，淋下調勻的麻辣汁便可上桌。

Ingredients

250g. fish fillet, 1 cucumber, 1~2 long bamboo shoot(jiao-bai bamboo shoot), 1/2C. shredded green onion, cilantro sections, green onion, ginger, wine

Seasonings

Ⓐa pinch of salt and pepper, 3T. water, 1T. cornstarch
Ⓑ1t. sesame seeds paste, 2T. soy sauce, 1/2T.sugar, 2T. water, 1/2T. sesame oil, 1t. vinegar, 1t. smashed garlic, 1/2t. brown peppercorn powder, 1/2T.red chili oil

Procedure

1. Cut fish fillet into thick strips. Mix with salt and water first, then add pepper and cornstarch in, mix evenly. Stay for 20 minutes.
2. Rinse cucumber, pat dry. Cut into thin slices, place on plate. Shred long bamboo shoot.
3. Blanch bamboo shoot, drain and place on top of cucumber. Add ginger, green onion and wine to water, blanch fish to done, remove, mix with green onion threads, put on top of bamboo shoot, sprinkle cilantro. Drizzle mixed seasoningsⒷ over fish.

軟溜魚帶粉

Sweet and Sour Fish with Bean Threads

材料

潮鯛魚肉 250 公克、蔥 2 支、薑 2 片、粉絲 1～2 把、酒 1 大匙、蝦米 2 大匙
蒜末 1 大匙、紅椒屑 1 大匙、蔥屑 1 大匙

調味料

Ⓐ鹽 1/4 茶匙、水 3 大匙、蛋白 1 大匙、太白粉 1 大匙
Ⓑ糖 3 大匙、醋 4 大匙、醬油 1 大匙、鹽 1/4 茶匙、太白粉 1 大匙、水 1½ 杯
胡椒粉少許、麻油 1/2 茶匙

做法

1. 魚肉先切除紅肉部分，再打斜切成片，沖洗一下擦乾。用調味料Ⓐ中的鹽和水抓拌，至產生黏性，加入蛋白，再拌勻後加入太白粉拌勻，醃 30 分鐘。蔥切段。
2. 鍋中煮滾 5 杯水，將粉絲燙熟撈出，放在大盤中。在水中加入蔥段、薑和酒 1 大匙，放入魚片以中小火燙煮至熟，撈出放在粉絲上。
3. 另用 2 大匙油炒香大蒜和蝦米屑等，倒入調味料Ⓑ煮滾，淋在魚肉上。

Ingredients

250g. fish fillet, sea bream,2 stalks green onion, 2 slices ginger, 2T. dried shrimp, 1T. chopped garlic, 1T. chopped red chili, 1T. chopped green onion, 1~2 bundles bean threads, 1T wine

Seasonings

Ⓐ1/4t. salt, 3T. water, 1T.egg white,1T. cornstarch
Ⓑ3T. sugar, 4T. vinegar, 1T. soy sauce, 1/4t. salt, 1½C. water, 1T. cornstarch, a pinch of pepper, 1/2t. sesame oil

Procedure

1. Remove red parts of fish, then slice it. Rinse with water,pat dry. Marinate with salt and water of seasoningsⒶ, then mix with egg white, when evenly ,mix with cornstarch,leave for 30 minutes. Cut green onion into sections.
2. Bring 5C. water to a boil, boil bean threads to done, drain and place on a plate. Add green onion, ginger and 1T. wine in water, blanch fish over medium low heat till done, remove, place on bean threads.
3. Heart 1T. oil to stir-fry all ingredients and seasoningsⒷ, when it boils, pour over fish.

沙茶拌魚條
Fish Strips with Sha-Cha Sauce

材料

魚肉 200 公克、西芹 1 支、水發木耳 1/2 杯
紅甜椒絲 1/2 杯、酒 1 大匙（燙魚用）

調味料

Ⓐ鹽少許、胡椒粉少許、太白粉 1 大匙
水 3 大匙
Ⓑ沙茶醬 1½ 大匙、糖 1/2 茶匙、水 3 大匙
醬油 1/2 大匙

做法

1. 魚肉切成如拇指般粗條，用調味料Ⓐ拌勻醃 20 分鐘。
2. 西芹撕去老筋，斜切成片狀。木耳泡軟並摘成小朵。
3. 煮滾一鍋水，放入木耳和西芹燙 1 分鐘，撈出後和紅甜椒一起放在碗中，用 1/2 量的沙茶醬拌勻，鋪放盤中。
4. 水中加酒，放入魚條燙熟，撈出也放碗中，加剩下的調味料拌勻亦裝入盤中。

Ingredients

200g. fish fillet, 1piece celery, 1/2C. soaked black fungus, 1/2C. shredded red chili, 1T. wine

Seasonings

Ⓐa pinch of salt and pepper, 3T. water, 1T cornstarch
Ⓑ1½T. sha-cha sauce, 1/2T. soy sauce, 1/2t. sugar, 3T. water

Procedure

1. Cut the fish fillet into the size about the thumb, mix with seasonings Ⓐ for 20 minutes.
2. Trim and slice celery. Trim the soaked fungus.
3. Boil fungus and celery for 1 minute, drain. Put them in a bowl with red chili shreds, mix half portion of seasoningsⒷ in, place on plate.
4. Add a little of wine into water, blanch fish strips, drain and mix with the remaining sha-cha sauce, place on top of vegetables.

豌豆米燴魚丁
Diced Fish with Snow Peas

材料
魚肉 150 公克、豌豆仁 150 公克、筍丁 2 大匙
竹笙 3 條、蔥花 1 大匙

調味料
Ⓐ鹽、胡椒粉、太白粉各少許、水 2 大匙
Ⓑ酒 1 茶匙、鹽 1/3 茶匙、白胡椒粉少許
清湯 1 杯、太白粉水適量、麻油少許

做法
1. 魚肉切成丁，用調味料Ⓐ拌勻，再醃約 10 分鐘。
2. 豌豆仁洗淨。竹笙泡軟，切成片。
3. 煮滾水 4 杯，放下豌豆仁和竹笙川燙 40～50 秒，撈出，再放入魚丁燙 10 秒鐘即撈出。
4. 起油鍋用 1 大匙油爆香蔥花，淋下酒，加入鹽、白胡椒粉和清湯煮滾，放入竹笙、筍丁和豌豆米，煮滾後放下魚丁再燴煮一滾，勾芡後滴下麻油拌勻，裝盤。

Ingredients
150g. fish fillet, 150g. snow beans, 2T. diced bamboo shoot, 3 pieces bamboo mushroom, 1T. chopped green onion

Seasonings
Ⓐa little of salt, pepper and cornstarch, 2T. water
Ⓑ1t. wine, 1/3t. salt, a pinch white pepper, 1C. soup stock, cornstarch paste, sesame oil

Procedure
1. Dice fish fillet, mix with seasoningsⒶ for 10 minutes.
2. Rinse peas. Soak bamboo mushroom to soft, then slice it.
3. Bring 4C. water to a boil, boil peas and mushroom for 40~50 seconds, drain. Boil diced fish for 10 seconds, drain.
4. Stir-fry green onion with 1T. oil, add 3 kinds of vegetables, sprinkle wine, add salt, white pepper and soup stock, cook until boils. Add bamboo mushroom ,diced bamboo shoot and snow beans, cook until boils. Add fish, thicken with cornstarch paste, drop sesame oil, remove to plate.

Fry 煎

在「煎」之前——

把外層表皮煎成金黃色、香香脆脆的，裡面魚肉還嫩嫩又有汁，是煎魚的最高境界！把一條魚煎的完整好看又好吃，應該是要練習幾次才會成功的，先來記住幾個重點：

1. 要先滑油盪鍋，旺火熱鍋後下油，讓油盪過鍋面，鍋子充分吸收油之後才會潤滑，煎時魚皮才不會黏住鍋底。使用不沾鍋則不需要盪鍋。
2. 盪鍋後另外加入適量的油。煎和炸不同，油不能多，通常是不超過材料厚度的一半。大火把油燒到熱，放入魚，大火先煎約 40 ～ 50 秒（魚小的話時間縮短），等表面黃了，汁封住不會流失，再改小火慢慢煎。魚最脆弱是在半生不熟和剛剛熟時，這時急著去翻面就一定會碎掉，所以要耐心等它煎硬、煎熟表皮時再翻面。
3. 翻面後也是大火先煎一下，等表面黃了，再改小火煎熟。出鍋前可以再用大火煎一下兩面，確定顏色和脆度。
4. 出鍋前用鏟子盛住魚，盡量將油滴掉，或在盤中先墊一張紙巾，吸掉一些油漬。

另外醃過的魚在下鍋前，要先擦乾水分，以免油爆。有些人喜歡沾一層粉料再炸（麵粉或太白粉），則要等下鍋前再沾，太早沾會使粉料吸入汁而潮濕，達不到乾香、酥脆的效果，這些也是要注意的細節。

魚煎過之後也可以再去烹、燴或溜，結合兩種烹調方法，使它具有另一種滋味，但是通常再烹調時仍會加調味料，因此醃魚時不要太鹹。

西餐中烹調魚，常會採煎的方法，煎香之後淋些魚高湯燜熟，再做個喜愛的醬汁搭配，這種加了水一起煎的方法，雖沒有脆脆的外皮，但是能保持魚肉的汁，使魚肉更嫩又較不油膩。後面食譜中有 3 道西式的做法可以參考。

同樣的，魚如果太大怕煎不熟，也可以在翻面後加少許水來燜煎，蓋上鍋蓋以使熱氣充足一些，比較容易熟。煎熟之後，再改大火煎脆表皮。喜歡更脆一點，可以再沿魚的周邊，加入少許油一起煎。

幾乎是任何魚都可以煎來吃，魚的鮮度差一點也不要緊。任何事都是，會了就不覺得難。會煎魚了，應該覺得煎一條好吃又好看的魚是一件很簡單的事了吧？

一般推薦可用於煎的魚種有

虱目魚 · 赤鯮魚 · 皇帝魚 · 金目鱸魚 · 鯧魚 · 帶魚 · 鱈魚

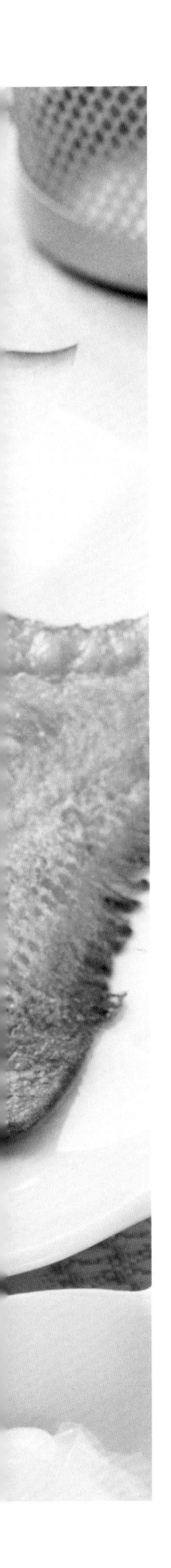

香煎鯧魚
Fried Pomfret

材料
鯧魚 1 條（約 1 斤）、蔥 2 支、薑 2 片

調味料
Ⓐ鹽 1 茶匙、酒 2 大匙、油 5 大匙
Ⓑ胡椒粉、美奶滋、花椒鹽等各適量

做法
1. 鯧魚修剪魚鰭和魚尾，在兩面魚肉上劃切斜而深的刀口。
2. 蔥和薑片拍碎，混合鹽和酒抹擦在魚的全身，醃約 30 分鐘。
3. 鍋中加熱 5 大匙油，油熱後放下鯧魚，大火先煎 1 分鐘，再改中小火煎熟，翻面再煎直到煎熟且酥黃。（煎魚的重點步驟請參考「在煎之前」。）
4. 盛入盤上，撒上胡椒粉、花椒鹽或附美奶滋，或其他喜愛的沾料上桌。

Ingredients
1 pomfret(about 600g.), 2stalks green onion, 2slices ginger

Seasonings
Ⓐ1t. salt, 2T. wine, 5T. oil
Ⓑpepper, mayonnaise, brown peppercorn salt

Procedure
1. Trim the fins and tail, score a few cuts on both sides.
2. Mix crushed green onion and ginger with salt and wine, rub fish with the mixture. Marinate for 30 minutes.
3. Heat 5T. oil, fry fish with high heat for the first minute, reduce to medium low heat until both sides done.
4. Remove to plate, sprinkle pepper on fish or serve with mayonnaise or brown peppercorn salt.

蒜烹皇帝魚
Fried Flounder with Garlic Sauce

材料
皇帝魚（比目魚）1 條、大蒜屑 1 大匙
麵粉 3 大匙、蔥屑 2 大匙、紅辣椒 1 支
青蒜絲少許

調味料
Ⓐ鹽 1/4 茶匙、胡椒粉 1/6 茶匙
Ⓑ酒 1 大匙、醬油 2 大匙、胡椒粉少許
糖 1 茶匙、麻油 1 茶匙、水 2/3 杯

做法
1. 皇帝魚打理乾淨，抹上調味料Ⓐ放置 10 分鐘後，魚身上拍上麵粉，用熱油煎至黃而酥脆，取出。紅辣椒切段。
2. 另用 2 大匙油爆香大蒜屑、蔥屑及紅辣椒段，淋下調味料 Ⓑ 煮滾。
3. 小心將魚放回鍋中，與湯汁一起煮滾，將湯汁多淋在魚身上，待汁被吸收，撒下青蒜絲，關火，盛入碟中。

Ingredients
1 flounder, 3T. flour, 1T. chopped garlic, 2T. chopped green onion, 1 red chili, green garlic shreds

Seasonings
Ⓐ1/4t. salt, 1/6t. pepper
Ⓑ1T. wine, 2T. soy sauce, 2/3C. water, 1t. sugar, 1t. sesame oil, a pinch of pepper

Procedure
1. Rinse fish. Sprinkle with seasoningsⒶ, leave for 10 minutes. Pat a flour on fish. Fry with hot oil until done. Cut red chili into sections.
2. Heat another 2T. oil to stir-fry garlic, green onion and red chili, pour seasoningsⒷ, bring the sauce to a boil.
3. Return flounder to wok, cook with sauce, pour sauce over fish while cooking. When sauce is absorbed, sprinkle green garlic shreds. Remove to plate.

XO 醬烹金線魚
Fried Fish with XO Sauce

材料
金線魚 1 條（約 450 公克）
西生菜絲 1 杯、洋蔥屑 1 大匙

調味料
Ⓐ鹽 1/2 茶匙、酒 1 大匙、胡椒粉少許
蔥段、薑片
ⒷXO 醬 2 大匙、水 4 大匙

做法
1. 魚打理乾淨後，切上 2～3 道刀口。蔥段和薑片略拍碎，混合調味料Ⓐ擦抹魚的兩面，放置 10 分鐘以上。
2. 鍋中燒熱 3 大匙油，放下金線魚煎熟且外層酥脆，盛放在生菜絲上。
3. 另用 1 大匙油炒香洋蔥屑，放入 XO 醬和水炒勻，淋在金線魚上。

Ingredients
1 fish(about 450g.), 1C. shredded lettuce, 1T. chopped onion

Seasonings
Ⓐ1/2t. salt, 1T. wine, a pinch pepper, green onion sections, ginger slices
Ⓑ2T. XO sauce, 4T. water

Procedure
1. Rinse fish. Make a few cuts on fish. Marinate fish with crushed green onion, ginger and seasoningsⒶ for 10 minutes.
2. Heat 3T. oil, fry fish until done, and the skin becomes brown and crispy. Remove to the top of shredded lettuce.
3. Stir-fry onion with 1T. oil, add seasoningsⒷ, mix and drizzle over fish.

軟煎鱸魚佐奶油醬汁

Fried Fish Fillet with Cream Sauce

材料

鱸魚 1 條、綠蘆筍 4 ～ 5 支、洋菇 5 粒、紅蔥屑 1 大匙、大蒜屑 1 茶匙
麵粉 3 大匙、蛋 1 個、巴西利末少許

調味料

Ⓐ鹽、白胡椒粉各適量、奶油 1 大匙
Ⓑ白酒 1/4 杯、魚高湯 1 杯、鮮奶油 1/4 杯、鹽適量、奶油 1 大匙

做法

1. 綠蘆筍切段；洋菇切片；鱸魚由背部下刀，片下兩邊魚肉，撒鹽和胡椒粉調味。沾麵粉後再沾上蛋汁，用奶油煎黃表面，淋下 4 大匙水，蓋上鍋蓋，改小火慢慢燜煎至熟。
2. 煎魚的同時，另用平底鍋熱 1 大匙油，小火炒香紅蔥屑和大蒜屑，加入洋菇片再炒一下，淋下白酒和魚高湯，煮至湯汁收縮到 1/2 杯。
3. 醬汁中加入鮮奶油、鹽和奶油攪勻，盛在盤中，再將煎好的魚排放在上面，撒下巴西利末。
4. 搭配炒過的綠蘆筍或其他青菜在盤邊。

魚高湯

將蔥段、薑片和魚頭、魚骨先用油煎香，淋 1 大匙白酒和水 3 杯，小火熬煮 20 分鐘，過濾即可。

Ingredients

1 sea bass, 4~5stalks asparagus, 5 mushrooms, 1T. chopped red shallot, 1t. chopped garlic, 3T. flour, 1egg, chopped parsley

Seasonings

Ⓐa pinch of salt and pepper, 1T. butter
Ⓑ1/4C. white wine, 1C. fish stock, 1/4C. cream, salt to taste, 1T. butter

Procedure

1. Cut asparagus into sections, slice mushrooms. Remove fish meat from both sides, sprinkle salt pepper to taste. Coat with flour, dip in beaten egg mixture, fry with butter until outside become brown. Add 4T. water, covered, fry over low heat until done.
2. While frying the fish, heat 1T. oil in pan, stir-fry red shallot and garlic till fragrant. Add sliced mushroom, stir-fry again, sprinkle wine and fish stock, cook until stock reduce to 1/2 cup.
3. Add cream, salt and butter, mix evenly, pour over the serving plate. Arrange on sauce, sprinkle some chopped parsley.
4. Serve with stir-fried asparagus or other vegetable you like.

To make fish stock

Stir-fry green onion, ginger, fish head and fish bones with 2T. oil, sprinkle white wine and 3 cups of water, boil for 20 minutes, drain off the bones.

鹹酥赤鯮
Fried Crispy Fish

材料
赤鯮魚或嘉臘魚 1 條、蔥屑 3 大匙
大蒜屑 1 大匙、紅辣椒屑 1 大匙

調味料：
Ⓐ鹽 2/3 茶匙、胡椒粉少許
Ⓑ醬油 2 大匙、酒 1 大匙、糖 1/2 大匙
水 1/3 杯、麻油 1/2 茶匙

做法
1. 將魚刮乾淨，在魚身兩面切上深且斜的刀口，撒下鹽抹勻，醃 10 分鐘。
2. 用約 6 大匙油煎至魚肉酥脆為止。盛在碟中，撒下胡椒粉少許。
3. 起油鍋炒香蔥、大蒜與紅辣椒屑，加入調味料Ⓑ，大火煮滾後馬上澆到魚身上面。

Ingredients
1 fish, 3T. chopped green onion, 1T. chopped garlic, 1T. chopped red chili

Seasonings
Ⓐ2/3t. salt , a pinch of pepper
Ⓑ2T. soy sauce, 1T. wine, 1/2T. sugar, 1/3C. water,1/2t. sesame oil

Procedure
1. Rinse fish. Score diagonally on both sides. Marinate with salt for 10 minutes.
2. Fry the fish with 6T. oil, when fish is done and the skin become crispy and brown. Sprinkle with salt and pepper.
3. Stir-fry green onion, garlic and red chili, add seasoningsⒷ, cook over high heat, pour over fish when sauce is boiled.

椒鹽虱目魚

Brown Peppercorn Salt Flavored Fish

材料

新鮮虱目魚中段 1 段、糖醋黃瓜 1/2 杯
花椒粒 2 大匙、麵粉 2 大匙

調味料

Ⓐ鹽 1 茶匙、酒 1 大匙
Ⓑ花椒鹽 2 茶匙

做法

1. 虱目魚取中段一段，也可以去掉大骨及內臟，剖開成為腹部相連之一大片。
2. 乾鍋中把花椒粒用小火慢慢炒香，撒在魚身上，再加鹽、酒抹勻，醃 30 分鐘。
3. 去掉花椒粒，在魚身拍上薄薄的乾麵粉，下鍋煎至黃且魚肉熟，上碟。可附上五香花椒鹽和糖醋黃瓜以配食。

糖醋黃瓜

黃瓜切片，用鹽醃 10 ～ 15 分鐘，擠乾水分。加糖和醋拌勻，放置 1 小時以上便可。

Ingredients

1 milk fish, 2T. brown peppercorn, 1/2C. sweet and sour cucumber, 2T. flour

Seasonings

Ⓐ1t.salt, 1T. wine,
Ⓑ2t. brown pepper salt

Procedure

1. Choose the center part of milk fish, you may cut it open and remove the spine.'
2. Stir-fry brown pepper corn in a wok without oil, when fragrant, add salt and wine, marinate fish for 30 minutes.
3. Pat a little of flour on fish, fry with hot oil over low heat. When the skin become brown and the meat is done, remove. Serve with brown pepper salt and sweet & sour cucumber.

To make sweet & sour cucumber

Mix sliced cucumber with a little of salt for 10~15minutes, rinse and squeeze out the water, mix with sugar and vinegar, leave for 1 hour.

香酥魚塔
Crispy Fish Pie

材料
白色魚肉 200 公克、蔥花 1 大匙、香菜 1 大匙、熟胡蘿蔔絲 1/3 杯
豆腐衣 2 張、荸薺 5 粒

調味料
薑汁 1/2 茶匙、鹽 1/3 茶匙、蛋白 1 大匙、太白粉 1 茶匙、麻油 1/2 茶匙

做法
1. 魚肉切成小丁，先加入調味料拌勻，再加入蔥花、香菜、荸薺（先剁碎並擠乾水分）和胡蘿蔔絲拌勻。
2. 兩張豆腐衣相疊，放上魚肉料成一長條形，包捲成寬扁的長條後，切成兩段。
3. 平底鍋中熱 4 ～ 5 大匙油，魚捲接口面朝下，放入鍋中，用小火煎熟。
4. 放在紙巾上略吸去油份，切成三角形，排入碟中。

Ingredients
200g. fish fillet, 1T. chopped green onion, 1T. chopped cilantro, 5 pieces water chestnut, 1/3C. cooked carrot, 2 pieces dried bean curd sheet

Seasonings
1/2t. ginger juice, 1/3t. salt, 1T. egg white, 1t. cornstarch, 1/2t. sesame oil

Procedure
1. Dice fish fillet, mix with seasonings, add all the chopped vegetables.
2. Double the dried bean curd sheets, place fish meat on top, roll it into a flat roll. Cut into 2 sections.
3. Heat 4T~ 5T.oil in a frying pan, place fish rolls on the pan with the edge side down. Fry over low heat till done.
4. Place on a paper tower to get rid off some oil, cut into triangles. Place on plate.

煎魚捲佐奶油醬汁
Fried Fish Rolls with Creamy Wine Sauce

材料
潮鯛魚片 300 公克、培根 2 片、高麗菜絲 1/2 杯、紅蔥頭 3 粒、洋蔥絲 1/2 杯
胡蘿蔔絲 1/4 杯、白胡椒粒 1 茶匙、奶油 1 大匙、烤熟腰果 5 ～ 6 粒

配菜
洋菇、新鮮香菇、紅甜椒丁各隨意

調味料
Ⓐ鹽、胡椒粉各適量
Ⓑ白酒 1/4 杯、魚高湯 1 杯、鮮奶油 1/4 杯、鹽、胡椒粉適量

做法
1. 魚肉要順紋切成薄片，撒上調味料Ⓐ。培根切絲。
2. 用 2 大匙油炒香培根絲，再加洋蔥絲、高麗菜絲和胡蘿蔔絲，炒到蔬菜料變軟，加少許鹽和水，煮約 2 ～ 3 分鐘。略勾芡後，盛出放涼。
3. 腰果或其他核果類均可，烤熟放涼後，略搗碎。
4. 魚肉鋪平，捲入適量的蔬菜料，做成魚捲。魚捲接頭處朝下，放入鍋中先煎 15 ～ 20 秒鐘至定型，倒入 1/2 杯水，蓋上鍋蓋小火煎熟。盛出裝盤。
5. 鍋中再加少許橄欖油，炒香紅蔥頭屑，加入白胡椒粒和調味料Ⓑ，煮至汁略濃稠，加入奶油攪勻，淋在魚捲上，再撒上腰果屑。
6. 配菜的菇類切適當大小，用油炒軟後加入紅甜椒，並加鹽和胡椒調味，拌勻裝盤。

Ingredients
300g. fish fillet, 2 slices bacon, 1/2C. onion shreds, 1/4C. carrot shreds, 1/2C. cabbage shreds, 3 cloves red shallot, 1t. white peppercorn , 1T. butter, 5~6 roasted cashew nuts

Side dish
fresh mushrooms, fresh black mushrooms, diced red bell pepper

Seasonings
Ⓐa pinch of salt and pepper
Ⓑ1/4C. white wine, 1C. fish stock, 1/4C. cream, salt, pepper

Procedure
1. Slice fish fillet into very thin slices. Sprinkle seasoningsⒶ. Shred bacon.
2. Stir-fry shredded bacon, add all vegetable shreds in, stir-fry until soft, add a little of salt and water, cook for 2~3 minutes. Thicken with cornstarch paste, remove and let cools.
3. Choose cashew nuts or other kinds of nuts, roast and crush slightly.
4. Roll the vegetables into fish slices. Place fish rolls on a frying pan (the edges on the bottom), fry for 15~20 seconds to make the rolls firmed. Add 1/2cup of water in, cover, fry over low heat until done. Remove to the serving plate.
5. Add a little of olive oil to pan, stir-fry chopped red shallot, add peppercorn and seasoningsⒷ, boil until the sauce become thick, mix cream in, pour over fish rolls, sprinkle nuts on top.
6. Stir-fry diced mushrooms, add red bell pepper in when mushrooms are soft, season with salt and pepper. Serve with fish rolls.

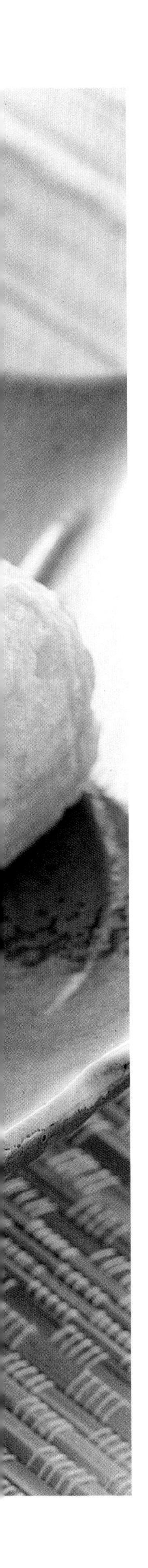

銀魚蛋捲
Egg Rolls with Small Fish

材料
魩仔魚 1/2 杯、蛋 5 個、韭菜 3 支

調味料
鹽適量、味醂 2 茶匙、太白粉 1 茶匙、水 2 茶匙

做法
1. 魩仔魚洗淨，瀝乾水分，用約 1 大匙油炒至乾香。
2. 蛋加調味料打散。韭菜切細小碎屑。
3. 小平底鍋燒熱，塗上一層油，倒下 1/3 量的蛋汁，攤開煎成蛋餅，見蛋汁將凝固至 7 分熟，從鍋邊捲起成筒狀。捲至最後，再倒下 1/3 量的蛋汁，撒下魩仔魚和韭菜屑，邊煎邊捲，捲完後再淋蛋汁，再捲起。
4. 做好蛋捲後改小火，慢慢煎熟。取出用壽司竹簾捲起定型，稍涼後切塊排盤。（可將魩仔魚改為煎熟的鮭魚）

Ingredients
1/2C. small white fish, 5 eggs, 3 stalks leek

Seasonings
salt, 2t. mirin, 1t. cornstarch, 2t. water

Procedure
1. Rinse and drain the small fish, stir-fry with a little of oil for a while.
2. Beat the eggs with seasonings until mixed. Cut leek thinly.
3. Heat the pan, brush a little of oil, pour 1/3 amount of egg mixture in, shake pan to make pan cake shape. When egg is almost solidify, roll it from one side. Pour another 1/3 of egg mixture in, add small fish and leek, fry and roll the egg mixture. Pour the rest egg in, roll into a cylinder.
4. Reduce heat to fry egg roll until done. Remove and place on a su-shi curtain, roll the curtain to shape the egg roll. Cut into pieces, serve. (You may use fried salmon instead of small fish.)

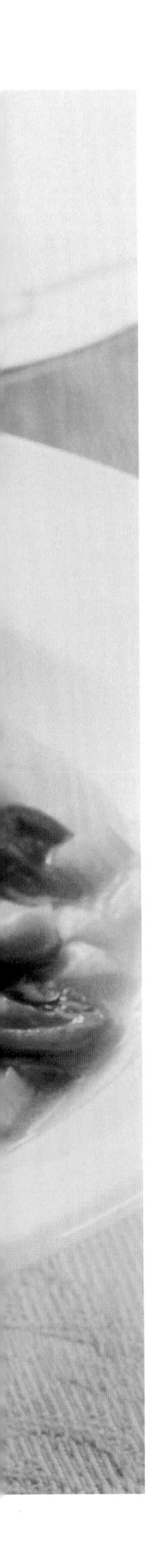

煎鯧魚佐蛤蜊香蔥醬汁

Fried Pomfret Fillet with Clam Sauce

材料

鯧魚 1 條、小蛤蜊 12 粒、西芹 1 支、蕃茄丁 2 大匙、松子 1 大匙
蔥屑 1 大匙、大蒜屑 1 大匙、蔥絲、香菜各少許

調味料

Ⓐ鹽、胡椒粉各少許、橄欖油 1 大匙、白酒 2 大匙
Ⓑ橄欖油 1/2 大匙、白醋數滴、鹽少許

做法

1. 將鯧魚由背部剖開，取下魚肉，撒少許鹽和胡椒粉醃 5 分鐘。平底鍋中熱橄欖油 1 大匙，放下魚肉（皮朝下）煎黃表面。翻面後淋下白酒，蓋上鍋蓋，改小火慢慢煎熟。
2. 蛤蜊加水 6 大匙，入鍋蒸至開口，剝出肉，同時留下湯汁約 150 cc。
3. 小鍋中放橄欖油、湯汁、蛤蜊肉和各種切丁料一起煮滾，加鹽調味，並滴下醋調勻，關火。
4. 醬汁盛盤底，上面放魚片，撒下松子，再放上蔥絲和香菜點綴。

Ingredients

1 pomfret, 12 clams, 1 piece celery, 2T. diced tomato, 1T. pine nuts, 1T. chopped green onion, 1T. chopped garlic, shredded green onion, cilantro

Seasonings

Ⓐslat and pepper to taste, 1T. olive oil, 2T. white wine
Ⓑ1/2T. olive oil, a few drops of white color vinegar, a pinch of salt

Procedure

1. Remove meat from pomfret. Sprinkle salt and pepper on fish, let it stay for 5 minutes. Heat 1T. olive oil in a frying pan, fry the fish with skin side down till light brown. Sprinkle white wine, turn to low heat, cover. Fry until done.
2. Steam clams and 6T. water until clam opened, remove meat, reserve 150cc juice.
3. Bring olive oil, clam stock, clams and diced vegetables to a boil, drop vinegar and salt, turn off the heat.
4. Place sauce on serving plate, arrange fish on, sprinkle pine nuts, garnish with green onion and cilantro. Serve.

Deep fry 炸

在「炸」之前——

「炸」是給魚做造型最好的方法。一條魚在切上花紋或去骨取肉之後，要利用炸來給它定型。有名的「糖醋魚」 如果沒有將魚肉一片片切上刀口，翻開來炸得酥酥脆脆的，即使淋上了糖醋汁也不會吸引人了。

魚肉在簡單的醃一下之後，無論是沾上粉或裹層麵糊去炸，都能把魚的鮮味留住，主要是因為油的高溫能迅速封住外層、並使內部魚肉很快致熟，保持住了魚肉的汁，（所以放涼了之後，風味就全沒了）。這種方法烹調出的魚的美妙滋味，是別的烹調法無法達到的。

要達到這種程度，首先要掌握油溫和火候大小。在炸的材料中，「魚」算是好炸的，魚肉容易熟，比較不會像大塊肉那樣外焦內生。通常沾了粉、裹了糊的魚下鍋時，油應該超過 8 分熱，以防止粉和糊脫落。下鍋之後，如果是改刀過的魚片、魚塊等不太大的或是切了刀口的，都可以用大火直接炸到表層酥脆，火力不夠大時，可以把魚撈出來，油燒熱後再放入炸第二次。

炸過的魚有的直接附沾料上桌沾食，沾料多以較爽口、帶酸酸口味、可以解油膩感的較受歡迎。有許多是另做調味汁淋上去的，或用具有香氣的材料烹一下增香，「奶油蒜香鯧魚」就是非常好吃的。

說到「炸」，雖然有許多人喜歡它的香脆，但卻因為怕留下一大鍋帶有魚腥味的炸油，不知該如何處理，而拒絕烹調它。其實一些裹了麵糊的魚塊，它的魚味是不會跑到油裡去的，而沾了乾粉（麵粉、太白粉、蕃薯粉、麵包粉）的魚塊，只要下鍋油溫夠熱，粉便會附著在魚上。至於少數散落在油裡的粉，待油冷卻後，自然沉落鍋底，把油倒入油罐中的時候，就不要了。

如果怕有魚味，可以在油中放 2 支蔥、2 片薑一起加熱，就可以除去氣味了。其實油炸最怕沾了醬油、又沒沾粉的材料，會使油裡帶有焦黑的油渣子。至於炸了之後會弄髒油的材料，可以用少量油分批來炸，炸過之後的油就不用了。

「炸」，可以豐富魚的變化，只要運用得當，就可以享受炸魚的美味。

一般推薦可用於炸的魚種有

沙梭魚 ・ 金線魚 ・ 鮭魚切片 ・ 鱈魚切片 ・ 黃魚 ・ 鯧魚 ・ 赤鯮魚 ・ 鮸魚
紅新娘

鮮果溜魚塊
Deep-fried Fish with Fruit Sauce

材料
石斑魚肉 250 公克、罐頭鳳梨 2 片
奇異果 1 個、芒果 1 個、紅甜椒 1/3 個
洋蔥丁 1 杯、太白粉 1 杯

調味料
Ⓐ鹽 1/4 茶匙、水 2 大匙、蛋黃 1 個
Ⓑ蕃茄醬 1 大匙、糖 2 大匙、醋 2 大匙
鹽 1/4 茶匙、太白粉 1 茶匙
罐頭鳳梨汁（可加水）1 杯

做法
1. 魚肉切成約 3 公分大小，用調味料Ⓐ拌勻醃 10 分鐘。
2. 各種水果和甜椒切成約 2 公分大小。
3. 魚肉沾裹上太白粉，投入 8 分熱油中炸熟且外層呈金黃色。
4. 用 2 湯匙油炒軟洋蔥，放下紅椒丁略炒，倒下調勻的調味料Ⓑ，煮滾後放下水果快速略拌合，關火再放入魚塊，一拌便可裝盤。

Ingredients
250g. grouper fish fillet, 2 slices canned pineapple, 1 kiwi, 1 mango, 1/3 red bell pepper, 1C. diced onion, 1C. cornstarch

Seasonings
Ⓐ1/4t. salt, 2T. water, 1 egg yolk
Ⓑ1T. ketchup, 2T. sugar, 2T. vinegar, 1/4t. salt, 1t. cornstarch, 1C. pine apple juice and water

Procedure
1. Cut fish fillet into 3cm cubes. Marinate with seasoningsⒶ for 10 minutes.
2. Cut all kinds of fruit and red pepper into 2cm pieces.
3. Coat fish with cornstarch. Deep-fry in 170°C oil until done, and the outside becomes golden brown. Drain.
4. Stir-fry onion, add red bell pepper and seasoningsⒷ, add fruits in when sauce boiled, mix quickly. Turn off the heat, add fish in, serve hot.

金沙鱈魚
Curry Cod with Popped Rice

材料
鱈魚 2 片（約 500 公克）、洋蔥屑 2 大匙
生鍋巴 3 片、紅椒丁 1 大匙、青椒丁 1 大匙

調味料
鹽 1/2 茶匙、蕃薯粉 1/2 杯、咖哩粉 1 大匙
胡椒粉少許

做法
1. 鱈魚剔除皮和骨，切成如大拇指般大小的條，撒上鹽和胡椒粉調味。放約 5 ～ 10 分鐘。沾上蕃薯粉。
2. 生鍋巴分成小塊，用熱油炸泡起，待稍涼，壓碎成小顆粒。如果是炸好的鍋巴，用手捏散成小顆粒。也可以用早餐的玉米片壓碎一些，要選沒有甜味的。
3. 待油降溫至 8 分熱，放入鱈魚條以中火炸熟，撈出。
4. 油倒出，放下洋蔥屑炒香，改小火，放入青、紅椒和鱈魚，撒下咖哩粉，同時用鏟子將鱈魚和咖哩粉拌勻，關火，撒下鍋巴，裝盤。

Ingredients
2 slices cod(about 500g.), 3 slices popped rice, 2T. chopped onion, 1T. diced red bell pepper, 1T. green pepper

Seasonings
1/2t. salt, a pinch of pepper, 1/2C. sweet potato powder, 1T. curry powder

Procedure
1. Remove skin and bones from cod. Cut into strips, sprinkle some salt and pepper to taste. Leave for 5~10 minutes.
2. Divide uncooked popped rice into small pieces, deep-fry with smoking hot oil over high heat, drain and crush. You may use the cereral instead of popped rice.
3. Lower the oil temperature to 160°C, deep-fry cod over medium heat until done, drain.
4. Pour away oil, stir-fry onion, when fragrant, reduce the heat, add green and red pepper and cod, sprinkle curry powder, mix evenly, turn off the heat, mix popped rice in, remove to serve.

酥炸魚條
Deep-fried Crispy Fish Strips

材料
新鮮魚肉 250 公克、麵粉 2 大匙

調味料
Ⓐ蛋白 1 大匙、鹽 1/2 茶匙、太白粉 1 茶匙、酒 2 茶匙、蔥 1 支（拍碎）
薑汁 1/2 茶匙

Ⓑ麵糊料：蛋 1 個、太白粉 3 大匙、低筋麵粉 3 大匙、鹽 1/4 茶匙
冰水酌量、油 1 大匙

做法
1. 魚肉切成如小拇指般粗細的長條。調味料Ⓐ先調勻，再將魚條放入拌合，醃 20 分鐘以上。
2. 蛋打散，依序加入麵糊料調勻成糊狀，最後落油拌勻。
3. 魚條撒上少許麵粉後拌入麵糊中。
4. 3 杯油燒至 8 分熱時改成小火，盡量分散魚條，一一投入油中。再開中火來炸。見魚條全部浮起，用筷子將黏在一起的魚條分開，再以大火炸 20 秒鐘，使外表脆硬便可撈出裝盤。附沾料上桌。

註：沾料可選用花椒鹽、蕃茄醬、沙拉醬、糖醋汁或蘿蔔泥柴魚汁。

Ingredients
250g. fish fillet, 2T. flour

Seasonings
1T. egg white, 1/2t. salt, 1t. cornstarch, 2t. wine, 1 stalk crushed green onion,1/2 ginger juice

Flour paste: 1 egg, 3T. cornstarch, 3T. flour, 1/4t. salt, ice water, 1T. oil

Procedure
1. Cut fish into strips(size about the little finger). Marinate with the mixed seasonings for 20 minutes.
2. Beat egg, add cornstarch, flour, salt and some water to make paste. Add oil at last.
3. Mix fish with 2T. flour, add into flour paste.
4. Heat 3 cups of oil to 160°C, deep-fry fish separately over low heat. Turn to medium heat, deep-fry until all flow up. Deep-fry over high heat for the last 20 seconds, drain. Serve with dipping sauce.

*You may choose brown peppercorn salt, or ketchup, or mayonnaise, or sweet & sour sauce, or Japanese style sauce as the dipping sauce.

橙汁魚排
Fish Fillet with Orange Sauce

材料
白色魚肉 250 公克、新鮮柳橙 1 個、蛋 1 個、麵粉 6 大匙、麵包粉 1 杯

調味料
Ⓐ鹽 1/2 茶匙、胡椒粉少許
Ⓑ瓶裝柳橙汁 1/2 杯、糖 2 大匙、檸檬汁 3 大匙、卡士達粉 1 茶匙
鹽 1/4 茶匙

做法
1. 魚肉打斜片切成薄片，撒上調味料Ⓐ醃 5 分鐘。
2. 1 個柳橙榨汁，約有 1/4 杯，再和其他調味料Ⓑ調勻（沒有新鮮柳橙時，可全用瓶裝的）。
3. 魚片依序沾上麵粉、蛋汁和麵包粉備炸。
4. 炸油燒至 8 分熱，放入魚片以中火炸熟，最後 15 秒改大火炸酥脆。撈出後瀝淨油漬。
5. 用 1 大匙油炒調味料Ⓑ，炒滾後關火，淋在魚片上。

Ingredients
250g. fish fillet, 1 fresh orange, 1egg, 6T. flour, 1 C. bread crumbs

Seasonings
Ⓐ1/2t. salt, a little of pepper
Ⓑ1/2C. orange juice in bottle, 2T. sugar, 3T. lemon juice, 1/4t. salt, 1t. custard powder

Procedure
1. Slice fish fillet, sprinkle seasoningsⒶ, stay for 5 minutes.
2. Squeeze the orange to get about 1/4 cup of fresh juice. Mix with other seasoningsⒷ (you may use all bottled orange juice)
3. Beat egg. Coat fish with flour first, then egg, and bread crumbs at last.
4. Deep-fry fish with 160°C oil over medium heat till done, turn to high heat for the last 15 seconds. Drain.
5. Stir-fry seasoningsⒷ, drizzle over fish when boils.

糖醋全魚
Fish with Sweet and Sour Sauce

材料
長型魚 1 條（約 1 斤重）、洋蔥丁 1/2 杯
蕃茄丁 1/2 杯、香菇丁 2 湯匙、青豆 2 湯匙
太白粉 1/2 杯

調味料
Ⓐ蔥 1 支、薑 2 片、鹽 2/3 茶匙、酒 1 大匙
Ⓑ蕃茄醬 3 大匙、糖 4 大匙、太白粉 1/2 大匙
鹽 1/4 茶匙、水 1/2 杯、麻油 1/2 茶匙
醋 4 大匙

做法
1. 魚打理乾淨，在兩側魚肉上，打斜刀切成深而薄的刀口，用調味料Ⓐ醃 15 分鐘。
2. 用太白粉沾裹魚身，投入熱油中炸兩次至酥而脆，撈出，瀝乾油。放在大盤中。
3. 用 2 大匙油先炒香洋蔥丁，再放入香菇丁和蕃茄丁，並將調勻的調味料Ⓑ倒入煮滾，放下青豆一拌，全部淋在魚身上。

Ingredients
1 long shaped fish(about 600g.), 1/2C. diced onion, 1/2C. diced tomato, 2 T. diced black mushroom, 2T. snow peas, 1/2C. cornstarch

Seasonings
Ⓐ1 stalk green onion, 2 slices ginger, 2/3t. salt, 1T. wine
Ⓑ3T. ketchup, 4T. sugar, 4T. vinegar, 1/4t. salt, 1/2C. water, 1/2T. cornstarch , 1/2t. sesame oil

Procedure
1. Rinse fish, score diagonally on both sides of fish. Marinate with seasoningsⒶ for 15 minutes.
2. Coat fish with cornstarch , deep-fry in hot oil twice to get the crispy outside. Place on serving plate.
3. Stir-fry onion, mushroom and tomato in 2T. oil, add seasoningsⒷ, bring to a boil, add snow peas, pour over fish.

奶油蒜香鯧魚球
Pomfret with Butter and Garlic

材料
鯧魚 1 條、大蒜片 1/3 杯、太白粉 5 大匙
奶油 1 大匙、檸檬 1/2 個

調味料
Ⓐ鹽 2/3 茶匙、酒 1 大匙、蔥 2 支、薑 3 片
Ⓑ鹽 1/4 茶匙、黑胡椒粉 1/6 茶匙、酒 2 茶匙

做法
1. 鯧魚剔下兩面魚肉，切成約 4 公分大小，和魚頭、魚骨一起用調味料Ⓐ拌勻醃 10 分鐘。
2. 魚肉等全部沾上太白粉後，先把魚骨炸脆排入盤中，再將魚肉炸至酥且金黃，撈出並瀝淨油。
3. 用 2 大匙油炒黃大蒜片（或入油鍋炸黃），放入奶油、調味料Ⓑ和魚塊，快速兜勻即可裝盤。附檸檬片上桌。

Ingredients
1 pomfret, 1/3C. sliced garlic, 5T. cornstarch, 1T. butter, 1/2 lemon,

Seasonings
Ⓐ2/3t. salt, 1T. wine, 2 stalks green onion, 3slices ginger
Ⓑ1/4t. salt, 1/6t. black pepper, 2t. wine

Procedure
1. Remove 2 pieces of meat from each side of pomfret, cut into 4cm pieces, place meat and bones in a bowl, marinate with seasoningsⒶ for 10 minutes.
2. Coat fish with cornstarch. Deep-fry bones first until crispy, drain and arrange on the serving plate. Deep-fry fish meat until brown and crispy, drain.
3. Heat 2T. oil to fry garlic, add butter, seasoningsⒷ and fish meat, mix evenly and quickly, place on top of bones. Serve with lemon slices.

滿載而歸

Fish with Treasures

材料
長型魚 1 條、魚肉 150 公克、魚丸 6 粒、紅甜椒 1/2 個、洋蔥丁 1/3 杯
綠蘆筍 3 支（燙過）、麵粉 2 大匙、太白粉 4 大匙、生菜絲 1 杯

醃魚料
蔥 1 支、薑 2 片、鹽 1/2 茶匙、酒 1 大匙、胡椒粉少許

調味料
蕃茄醬 2 大匙、糖 3 大匙、醋 3 大匙、鹽 1/4 茶匙、水 1 杯、麻油少許

做法
1. 魚打理乾淨後由背部剖開，沿著魚骨兩側片切開，用剪刀剪除魚大骨，成為魚腹部肉仍相連的船形，用醃魚料醃 10 ～ 15 分鐘。
2. 魚肉切成小塊，加入魚中一起醃。魚丸切半，燙過並沖涼的綠蘆筍和紅甜椒分別切成小塊。調味料在碗中調好。
3. 混合麵粉和太白粉，沾滿魚身內外和魚肉，用熱油把魚炸的又酥且香，同時保持魚身炸成凹船形，放在盤中生菜絲上。魚肉炸熟撈出。
4. 起油鍋中爆香洋蔥丁，放入各種配料和調味料，煮滾後拌入魚肉，一起盛放在魚身上。

Ingredients
1 long shaped fish, 150g. fish fillet, 6 pieces fish balls, 1/2 red bell pepper, 1/3C. onion, 3 asparagus(boiled), 2T. flour, 4T. cornstarch, 1C. shredded lettuce

Seasonings
Ⓐ1 stalk green onion, 2 slices ginger, 1/2t. salt, 1T. wine, pepper
Ⓑ2T. ketchup, 3T. sugar, 3T. vinegar, 1/4t. salt, 1C. water, sesame oil

Procedure
1. Rinse fish, score from the back, make the fish openly along the spine, cut off spine by scissors, make the fish look like a boat. Marinate with seasoningsⒶ for 10~15 minutes.
2. Cut fish fillet into pieces, marinate with the fish. Halve every fish ball. Cut asparagus and red bell pepper. Mix seasoningsBin a bowl.
3. Mix flour and cornstarch, coat fish and fillet with that powder. Deep-fry fish in hot oil until crispy. Drain, place on top of shredded lettuce. Deep-fry fish fillet, drain.
4. Stir-fry diced onion, add all ingredients and seasoningsⒷ, mix fish fillet in when the sauce boils. Remove to the center of the fish.

糖醋松鼠魚
Sweet and Sour Boneless Fish

材料
大黃魚（或其他長型魚）1 條（約 750 公克）、香菇丁 2 大匙、蕃茄丁 1/2 杯
洋蔥丁 1/2 杯 、青豆 2 大匙

調味料
Ⓐ雞蛋 1 個、麵粉 4 大匙、太白粉 4 大匙、水 4 大匙
Ⓑ醬油 1/2 大匙、酒 1 大匙、糖、白醋、蕃茄醬各 4 大匙、清水 6 大匙
太白粉 2 茶匙、鹽 1/2 茶匙、麻油 1 茶匙

做法
1. 先將魚頭切下，剖開成一大片後撒上少許鹽。將調味料Ⓐ調成麵糊。
2. 魚身部分剔除大骨，取下兩面魚肉，在魚肉上先直切 2 長刀，再在橫面每隔 1.5 公分切劃一刀。切妥之後，撒下鹽 1/2 茶匙以及酒 1 大匙醃 10 分鐘。
3. 炸油燒熱，投下沾裹了麵糊之魚頭和魚肉，用大火炸熟（約 3、4 分鐘），至十分酥脆時，撈出排盤中。
4. 鍋內用 2 大匙油先炒香洋蔥丁，再放入香菇丁、蕃茄丁、青豆與調味料Ⓑ，大火滾煮後澆到盤中之魚上。

Ingredients
1 long shape fish(about 750g.), 2T. diced black mushroom, 1/2C. diced tomato, 1/2C. diced onion, 2T. snow peas

Seasonings
Ⓐ1egg, 4T. flour, 4T. cornstarch, 4T. water
Ⓑ1/2T. soy sauce, 1T. wine, 4T. each of sugar, vinegar and ketchup, 6T. water, 2t. cornstarch, 1/2t. salt, 1t.sesame oil

Procedure
1. Rinse fish. Cut the head off and halve, sprinkle some salt on.
2. Cut off two pieces of meat, discard the spine. Score 2 straight cuts first, then every 1.5cm, make a vertical cut. Marinate with 1/2t. salt and 1T. wine for 10 minutes.
3. Coat fish head and 2 pieces of fish meat with mixed seasoningsⒶ. Deep-fry in hot oil for about 3~4 minutes, when it becomes crispy, drain.
4. Heat 2T. oil to stir-fry onion first, add other ingredients and seasoningsⒷ, bring to a boil, pour over fish.

雙味魚捲
Two Flavored Fish Rolls

材料
白色魚肉 300 公克、熟筍絲 1/3 杯、豆腐衣 4 張、韭菜屑 1/2 杯

調味料
Ⓐ鹽 1/2 茶匙、蛋白 1 大匙、胡椒粉 1/6 茶匙、蔥屑 1/2 大匙、麻油 1 茶匙
薑汁 1/2 茶匙、油 1 大匙
Ⓑ蒜屑 1/2 大匙、蔥花 1 大匙、糖 2 大匙、醋 2 大匙、水 4 大匙、麻油少許
鹽 1/4 茶匙、蕃茄醬 2 大匙、太白粉少許
Ⓒ花椒粉 1/2 茶匙、鹽 1 茶匙混合均勻

做法
1. 魚肉切成指甲大小片狀。大碗中將調味料Ⓐ調勻，再放入魚肉拌勻，醃約 10 分鐘。拌入筍絲和韭菜屑。
2. 豆腐衣每張切成 3 小張，包入適量的魚肉，包捲成小春捲形。
3. 用 1 大匙油爆香蒜屑和蔥花，再加入其它的調味料Ⓑ煮滾，做成甜酸汁，盛裝小碟中。
4. 鍋中炸油燒至 8 分熱，放入魚捲小火炸約 1 分鐘，最後改大火炸成金黃色，撈出，瀝乾油份，裝盤。附上花椒鹽和甜酸汁上桌。

Ingredients
300g. fish fillet, 1/3C. shredded bamboo shoot(cooked),
1/2C. diced leek, 4 pieces dried bean curd sheet

Seasonings
Ⓐ 1/2t. salt, 1T. egg white, 1T. oil, 1t. sesame oil, 1/6t. pepper, 1/2T. chopped green onion, 1/2t. ginger juice
Ⓑ 1/2T. chopped garlic, 1T. chopped green onion, 2T. ketchup, 2T. sugar, 2T. vinegar, 4T. water, 1/4t. salt, cornstarch, sesame oil
Ⓒ Mix 1/2t. paprika and 1t. salt well.

Procedure
1. Cut fish into small pieces. Marinate with mixed seasoningsⒶ for 10 minutes. Add bamboo shoot and leek.
2. Divide each bean curd sheet into 3 small pieces, wrap fish mixture into small pack.
3. Stir-fry garlic and green onion with 1T. oil, add other seasoningsⒷ to make the sweet & sour sauce.
4. Deep-fry fish rolls with 160°C oil for about 1 minute over low heat, turn to high heat at last 20 seconds. Drain, serve with brown pepper salt and sweet & sour sauce.

生汁芝麻魚
Deep-fried Fish with Mayonnaise

材料
魚肉 450 公克、罐頭鳳梨 4 片、青豆 2 大匙、玉米粉 1 杯、美奶滋 3 大匙
白芝麻 1 大匙、大芋頭 300 公克、生菜絲適量

調味料
鹽 1/3 茶匙、酒 1/2 茶匙、蛋白 1 大匙、太白粉 1 茶匙

做法
1. 魚肉切成 3 公分大小，用調味料拌勻，醃 5 分鐘。
2. 鳳梨片一切為四小片。
3. 芋頭切成細絲，在薄鹽水中漂洗一下，擦乾水分。拌上 2 大匙玉米粉，排在沾了油的（以免黏住芋頭）漏勺中，上面再壓一個小漏勺。一起放入油鍋中炸成有深度的鳥巢形，放在鋪了生菜絲的盤中。
4. 魚塊沾裹上玉米粉，投入熱油中炸至黃且酥脆，撈出。鳳梨片和青豆也在熱油中快速撈一下。
5. 魚塊和鳳梨、青豆同放碗中，擠上美奶滋快速一拌，裝入芋巢之中，撒下炒香的白芝麻。

Ingredients
450g. fish fillet, 4slices caned pineapple, 2T. snow peas, 1C. cornstarch, 3T. mayonnaise, 1T. sesame seeds, 300g. taro, shredded lettuce

Seasonings
1/3t. salt, 1/2t. wine, 1T. egg white, 1t. cornstarch

Procedure
1. Cut fish fillet into 3cm pieces. Mix with seasonings, marinate for 5minutes.
2. Cut pineapple into small pieces.
3. Shred taro, rinse with salty water, drain and pat dry. Mix with 2T. cornstarch, arrange on a strainer, press the top with another strainer. Deep-fry in smoking hot oil to form a bird nest shape. Remove and place on the shredded lettuce.
4. Coat fish with cornstarch, deep-fry until done, drain. Run the pineapple and snow peas through oil, drain. Place into a bowl.
5. Add mayonnaise in, mix quickly and evenly, remove to the nest, sprinkle stir-fried sesame seeds. Serve.

西炸鮭魚球

Deep-fried Salmon Balls

材料

新鮮鮭魚 120 公克、馬鈴薯 450 公克
洋蔥屑 1/2 杯、麵粉 3 大匙、蛋 1 個
麵包粉 1½ 杯

調味料

鹽 1/2 茶匙、胡椒粉少許

做法

1. 馬鈴薯煮軟透後取出，待稍涼後剝去外皮、壓成泥，放在大碗中。
2. 鮭魚抹少許鹽後入鍋蒸熟，剝成小粒。用 2 大匙油炒軟洋蔥屑，連魚肉、調味料一起和薯泥仔細拌勻，再分成小粒，搓成圓形。
3. 鮭魚球先沾一層麵粉，再沾上蛋汁，最後滾滿麵包粉，投入熱油中以中火炸黃，瀝乾油裝入盤中，可另附沙拉醬或蕃茄醬沾食。

Ingredients

120g. salmon, 450g. potato, 1/2C. chopped onion, 3T. flour, 1 egg, 1½C. bread crumbs

Seasonings

1/2t. salt, a pinch of salt

Procedure

1. Boil potato to soft. Peel and smash when it becomes cooler, place in a big bowl.
2. Rub salmon with some salt. Steam and separate into small pieces. Stir-fry chopped onion to soft, add salmon, seasonings and potato, mix evenly. Divide to small portions, form to a ball shape.
3. Coat salmon balls with flour first, dip in beaten egg, and cover with bread crumbs at last. Deep-fry in 180°C oil over medium heat until brown and crispy, drain. Place on plate, serve with dipping sauce, such as ketchup or mayonnaise or Worcestershire sauce.

紅糟魚條
Deep-fried Fish Strips

材料
潮鯛魚肉或海鰻 250 公克
蕃薯粉 2/3 杯

調味料
紅糟 2 大匙、糖 1 茶匙、鹽 1/3 茶匙
酒 1 大匙、薑汁 1 茶匙、胡椒粉少許
水 1 大匙、太白粉 1 大匙

做法
1. 將魚肉切成約 3 ～ 4 公分的粗條，拌上調勻的紅糟料（調味料），醃 20 分鐘。
2. 魚條沾裹上蕃薯粉。
3. 炸油燒至 8 分熱，將魚條一條一條投入，炸至脆硬即可撈出，裝盤。

Ingredients
250g. fish fillet, 2/3C. sweet potato powder

Seasonings
2T. fermented red wine lees, 1t. sugar, 1/3t. salt, 1T. wine, 1t. ginger juice, 1T. water, pepper, 1T. cornstarch

Procedure
1. Cut fish fillet into 3~4cm long strips, marinate with mixed seasonings for 20 minutes.
2. Coat fish with sweet potato powder.
3. Heat the oil to 160°C, drop fish strips one by one, deep-fry over medium heat until outside becomes crispy. Drain. Place on a plate.

雙菇燜鮮魚
Fish with Two Mushrooms

材料
白色魚肉 250 公克、新鮮香菇 3 ～ 4 朵、杏鮑菇 3 ～ 4 支
蔥段數支、薑片酌量

醃魚料
鹽 1/4 茶匙、糖 1/2 茶匙、胡椒粉少許、水 2 大匙、蛋白 1 大匙

蛋麵糊
蛋 1 個、麵粉 4 大匙、水酌量

調味料
醬油 2 大匙、鹽 1/4 茶匙、糖 1/4 茶匙、清湯或水 2/3 杯

做法
1. 魚肉打斜切片，用醃魚料醃 5 ～ 10 分鐘。沾上一層乾麵粉後再裹上蛋麵糊，入油鍋炸黃，撈出。
2. 香菇切厚片。杏鮑菇斜切片。
3. 用 2 大匙熱油炒香蔥段、薑片，再放入兩種菇類慢慢炒至香氣透出，淋下調味料煮滾半分鐘。
4. 再放下魚塊一起燜煮 1 ～ 2 分鐘，待吸收了滋味即可關火裝盤。

Ingredients
250g. fish fillet, 3~4 pieces fresh black mushroom, 3~4 pieces oyster mushroom, green onion sections, ginger slices

Seasonings
Ⓐ1/4t. salt, 1/2t. sugar, a pinch of pepper, 2T. water, 1T. egg white
Ⓑ1 egg, 4T. flour, water,
Ⓒ2T soy sauce, 1/4t. salt, 1/4t. sugar, 2/3C. soup stock or 2/3C. water

Procedure
1. Slice fish, marinate with seasoningsⒶ for 5~10 minutes. Coat the fish with a thin layer of flour first, then cover with mixed seasoningsⒷ. Deep-fry in hot oil till brown, drain.
2. Slice both mushrooms.
3. Stir-fry green onion and ginger with 2T. oil, add mushrooms in, stir-fry until fragrant, add seasonings Ⓒ, boil for 1/2minutes.
4. Add fish in, cook with mushrooms for 1~2 minutes, when fish absorb the sauce, turn off the heat and serve.

Bake 烤

在「烤」之前——

烤是人類最原始的烹調法，將材料經過醃漬入味後，架在鐵網上或放入烤爐中，用炭或柴加熱烤熟。「烤」因為用具不同而分明爐烤和暗爐烤。用炭火來烤雖然很香卻不能常吃，現在的烤多半是用烤箱來烤。

在現代家庭中，烤箱雖然普遍，但一般家庭用小烤箱的火力強度和均勻度，卻遠不如專業的，因此常有人抱怨效果不是很好，但是用在烤魚方面卻是綽綽有餘的。

因為魚肉容易熟，因此做烤魚的菜式，幾乎全是大火就可以了。如果想保持魚肉嫩，可以用鋁箔紙包著烤，或用鋁箔紙覆蓋在烤碗上來烤。紙包中加些水或高湯，藉著水氣的滋潤，可以使魚肉嫩又鮮美。

如果想要烤出來的魚有香香的焦痕，就不用覆蓋、直接烤。但是如果是醃過醬料或塗有醬汁的魚，在預熱烤箱時，雖可將溫度定在高溫 230 ～ 250℃，但在烤了 2 ～ 3 分鐘後就要把溫度降下來，以免太焦會有苦味。日式烤魚中有 3 種有名的烤法：鹽燒、照燒和味噌燒，都是以烤得香香的取勝，在餐廳中多是以直火來烤效果好，在家以烤箱來烤，火力掌握好的話，效果也不差。

烤魚時還要注意塗油，無論是用鋁箔紙或是用烤盤、烤架，都要塗上油，以免黏住不好翻面。

為了快速封住表面，保持肉質中的汁不流失，在烤之前也常用大火把兩面煎一下，再放在塗了少許油的烤盤上，送進烤箱烤熟。同樣的可以加少許水，以蒸烤的方式進行，使肉質較嫩；或直接烤，得到較焦香的口感，最後再淋上喜愛的醬汁。

早年香港西餐廳開始以西式的奶油糊加料，盛入烤碗後撒上起司粉，再烘烤出來的菜式叫做 「焗」，現在也有人稱 「焗烤」。這種焗烤的方法很適合烤魚，用奶油糊包住魚，使魚肉的原汁不會流失，魚肉也因為不直接接觸火而特別嫩。

「烤」既沒有油煙、又可以賦予魚特殊的香氣，何不試試烤魚？

一般推薦可用於 烤的魚種有

鮭魚頭 ・ 香魚 ・ 鱈魚 ・ 青衣 ・ 小黃魚 ・ 赤鯮魚 ・ 鯧魚 ・ 馬加魚切片 ・ 嘉鱲魚
皇帝魚

起司焗鮮魚
Baked Fish with Cheese

材料
圓鱈魚肉 300 公克、洋菇 6 粒、洋蔥丁 2 大匙、奶油 1 小塊、麵粉 4 大匙
青花菜 1 小棵、水或清湯 2½ 杯、Parmesan 起司粉 1 ～ 2 大匙
鮮奶油 2 大匙、披薩起司 1 ～ 2 大匙、土司麵包 2 ～ 3 片

調味料
Ⓐ鹽 1/4 茶匙、胡椒粉少許、太白粉 1 大匙
Ⓑ鹽、胡椒粉各適量調味

做法
1. 鱈魚去皮去骨後，將魚肉斜切成厚片，拌上調味料Ⓐ，醃約 20 分鐘。青花菜分成小朵，燙熟、沖涼。洋菇切片。土司麵包烤黃，切成小片。
2. 煮滾 4 杯水，放入魚片，小火燙 10 秒鐘左右即撈出。
3. 燒熱 3 大匙油，炒香洋蔥和洋菇，加入麵粉炒黃。慢慢加入清湯，攪拌成均勻的糊狀，加鹽和胡椒粉調味後拌入奶油和鮮奶油調勻，關火。
4. 烤碗底部鋪放土司片和少許麵糊料，再將魚片和青花菜全部裝好，蓋上剩餘的麵糊，撒下起司粉和披薩起司。
5. 烤箱預熱至 250℃，放入烤碗，用上火烤至起司融化且呈金黃色（約 12 ～ 18 分鐘）。取出趁熱食用。

Ingredients
300g. fish fillet, 6 pieces mushroom, 2T. chopped onion, 1 broccoli,
1 piece butter, 4T. flour, 2½C. soup stock or water, 2T. cream, 1~2T. Parmesan cheese,
1~2T. Pizza cheese, 2~3 pieces bread

Seasonings
Ⓐ1/4t. salt,a pinch of pepper,1T.cornstarch
ⒷSalt and pepper to taste

Procedure
1. Remove bones and skin from cod, slice fish fillet into thick pieces, mix with seasoningsⒶ, marinate for 20 minutes. Trim broccoli, blanch and rinse with cold water. Slice mushrooms. Toast the bread to brown, divide into small pieces.
2. Boil fish with boiling water for 10 seconds, drain.
3. Heat 3T. oil to stir-fry onion and mushrooms, add flour, stir-fry until light brown, add soup stock in, stir evenly to form a thick paste. Season with salt and pepper, mix with butter and cream.
4. Place toast and flour paste on the bottom of a baking ware, arrange fish and broccoli in, cover with remaining paste. Sprinkle two kinds of cheese.
5. Preheat oven to 250°C, bake until the surface become golden brown(about 12~18 minutes).

香蒜培根烤魚排
Fish Fillet with Bacon and Garlic

材料
鱈魚或白色魚肉 250 公克、金針菇 1 包
培根 4 片、大蒜屑 1 大匙、奶油 1 大匙
水 2 大匙

調味料
鹽、胡椒粉適量

做法
1. 鱈魚去皮去骨後撒少許鹽和胡椒粉，放置 5 ～ 10 分鐘。培根切段。
2. 金針菇洗淨，切除根部，散放在烤盤中（或用鋁箔紙折成烤盤狀），撒少許鹽和水 2 大匙。上面放上鱈魚。
3. 起油鍋炒香培根段和大蒜屑，盛放在鱈魚上。奶油分成小丁，也散放在魚上面。
4. 烤箱預熱到 250℃，放入魚排烤熟。

Ingredients
250g. cod or other kind of fish fillet, 1 pack needle mushroom, 4 slices bacon, 1T. chopped garlic, 1T.butter, 2T. water

Seasonings
salt and pepper to taste

Procedure
1. Remove skin and bones from fish, cut into big pieces. Sprinkle salt and pepper, leave for 5~10 minutes.Cut bacon into sections.
2. Trim and rinse needle mushroom, place on the baking ware(or use aluminum foil to make a plate). Sprinkle some salt and 2T. water. Place fish on top.
3. Stir-fry bacon slices and garlic, put on top of fish. Also add some butter cubes on top.
4. Preheat oven to 250°C, bake fish until done.

香蔥烤鮭魚

Baked Salmon with Green Onion

材料

鮭魚 1 片、蔥花 1/2 杯、鋁箔紙 1 張

調味料

Ⓐ鹽、胡椒粉各少許、油 1 大匙
Ⓑ美極鮮醬油 1/2 大匙、水 2 大匙
油 1/2 大匙

做法

1. 鮭魚連皮帶骨一起切成約 4 公分的塊狀，拌上調味料Ⓐ。蔥花中也拌上調味料Ⓑ。
2. 鋁箔紙裁成適當大小（或用烤碗，碗上蓋鋁箔紙），撒下 1/2 的蔥花，放上鮭魚塊，再撒上蔥花，包好鋁箔紙。
3. 烤箱預熱至 250℃，放入鮭魚烤大約 18 分鐘，即可取出。

註：喜歡魚肉烤得有焦痕的話，可以在 8 分鐘後打開鋁箔紙，再烤 6～7 分鐘。

Ingredients

1slice salmon, 1/2C. chopped green onion, a piece of aluminum foil

Seasonings

Ⓐa pinch of salt and pepper, 1T. oil
Ⓑ1/2T. soy sauce, 2T. water, 1/2T. oil

Procedure

1. Cut salmon into 4cm pieces with the bone and skin. Mix with seasoningsⒶ. Mix green onion with seasoningsⒷ.
2. Doubled the aluminum foil(or you may use a baking ware, covered by the foil), place half of the green onion on the bottom, put salmon on top, sprinkle the rest of the green onion on salmon. Pack the foil.
3. Preheat the oven to 250°C, bake for 18 minutes, remove.

*If you prefer to have some brown uu on salmon, you may open the foil after 8 minutes, then bake for another 6~7 minutes.

Cour Figeac

酒烤紙包魚

Baked Cod with White Wine Sauce

材料

鱈魚 1 大塊（約 5～6 公分寬）、新鮮香菇 4 朵、西芹 1 支、胡蘿蔔 1 小段
蔥 1 支、鋁箔紙 1 大張

調味料

Ⓐ鹽、胡椒粉各少許、白酒 2 茶匙
Ⓑ紅蔥頭屑 2 大匙、大蒜屑 2 茶匙、白酒 1/2 杯、高湯 3/4 杯、鮮奶油 2 大匙
鹽 1/3 茶匙、白胡椒粉少許、奶油 1 大匙

做法

1. 鱈魚剔下魚皮和骨，可以取下兩片魚肉。在魚肉上撒上少許鹽和胡椒粉。
2. 各種蔬菜料洗淨切片，蔥切絲。起油鍋，用 2 大匙油炒蔥絲和蔬菜料，加少許鹽和胡椒粉調味。
3. 鋁箔紙上舖上蔬菜料，放上鱈魚塊，滴上少許白酒，包好紙包。
4. 烤箱預熱至 250℃，放入紙包魚，烤 15 分鐘左右至魚已熟，取出。
5. 用油炒香紅蔥頭屑和大蒜屑，淋下白酒煮至酒精揮發，加入高湯煮滾，濾除碎屑，再加鮮奶油濃縮湯汁，用鹽和胡椒粉調味，最後再加奶油攪均勻。
6. 打開紙包魚，將蔬菜料和魚肉移到盤中，淋上香酒醬汁即可。

註：若放入烤碗中直接烤，應用鋁箔紙覆蓋，以保持魚肉的嫩度。

Ingredients

1 big piece of cod,about5~6cm wide, 4 pieces fresh black mushroom, 1 piece celery, 1 piece carrot, 1 stalk green onion, 1 piece aluminum foil

Seasonings

Ⓐsalt and pepper to taste, 2t. white wine
Ⓑ2T. chopped red shallot, 2t. chopped garlic, 1/2C. white wine, 3/4C. soup stock, 2T. cream, 1/3t. salt, a pinch of pepper, 1T. butter

Procedure

1. Remove skin and bones from cod, you may get 2 pieces fillet. Sprinkle some salt and pepper on fish.
2. Rinse and slice all vegetables, shred green onion. Stir-fry them with 2T. oil, season with salt and pepper.
3. Place vegetables on aluminum foil, put cod on top, drizzle 2T. white wine over fish. Fold and wrap into package.
4. Preheat oven to 250°C, bake fish for about 15 minutes until done. Remove.
5. Stir-fry red shallot and garlic with oil, add white wine, cook for 1minute. Add soup stock, bring to a boil, drain red shallot out. Mix cream into sauce, season with salt and pepper, add butter at last.
6. Remove vegetable and fish to the serving plate if prefer , drizzle sauce over fish.

* If you use a baking ware to bake this dish, you should cover with aluminum foil, to keep cod tender.

烤味噌魚片
Baked Miso Fish

材料
馬加魚肉（土魠魚）或旗魚 1 片
檸檬 1/2 個

調味料
味噌 1/2 杯、甜酒釀 1/2 杯、糖 1 大匙
甜酒釀汁 1/4 杯

做法
1. 酒釀捏碎和湯汁、味噌、糖一起攪拌均勻。
2. 魚快速沖洗一下，擦乾水分，放入味噌醃料中醃 6 小時以上。
3. 將魚快速沖一下，即放在塗油的烤架上。烤箱預熱至 240℃，放入味噌魚片，烤 2 分鐘後，改 180℃烤至熟（約需烤 15 ～ 18 分鐘）。
4. 取出放入盤中，附檸檬片上桌。

Ingredients
1 piece Chinese mackerel or other boneless fish, 1/2 lemon

Seasonings
1/2C. miso, 1/2C. fermented sweet wined rice, juice from wined rice, 1T. sugar

Procedure
1. Mix sweet wined rice well with the juice, miso and sugar.
2. Rinse fish quickly, pat dry. Marinate with miso sauce for 6 hour.
3. Rinse mackerel, place on the greased rack. Preheat oven to 240°C, bake fish for 2 minutes. After 2 minutes keep the temperature at 180°C until fish is done. (For about 15~18minutes)
4. Remove to serving plate, serve with lemon slices.

日式照燒魚
Baked Pomfret with Tariyaki Sauce

材料
鯧魚 1 條、白芝麻 1 小匙、檸檬 1/2 個或白蘿蔔泥適量

調味料
醬油 3 大匙、糖 2 大匙、酒 1 大匙
味醂 1 大匙，準備 2 份

做法
1. 將魚切大斜片，用調味料拌醃 30 分鐘，醃時要時常翻面。
2. 烤盤上鋪鋁箔紙，刷油後排上魚片。烤箱預熱後，放入魚排，用大火烤熟（需要翻面）。
3. 將另一份調味料煮成濃汁，在烤魚的時候，把汁往魚上塗刷數次，見魚的表面已成褐色，便可取出，撒下炒過之白芝麻，附上檸檬片或蘿蔔泥。

Ingredients
1 pomfret, 1t. white sesame seeds, 1/2 lemon or mashed radish

Seasonings
3T. soy sauce, 2T. sugar, 1T. wine, 1T. mirin

Procedure
1. Slice fish into big pieces, marinate with 30 minutes, turn the fish over and over while marinate.
2. Place aluminum foil on baking try, brush some oil on foil, arrange fish on. Preheat oven to 240°C, bake fish over high heat until done.
3. Cook another potion of seasonings over low until sauce becomes sticky, brush the sauce over fish while baking. Remove when fish is done and brown. Remove and sprinkle stir-fried sesame seeds on. Serve with lemon slice or smashed radish.

鹽烤香魚

Baked Smelt, Japanese Style

材料

香魚 1 尾

調味料

酒 1 茶匙、鹽 1/2 大匙

做法

1. 香魚去鰓沖洗一下，擦乾水分，在魚身兩面平均撒下少許的鹽和酒，用手指抹勻。手上再捏少許鹽粒，沾在魚之背鰭和尾鰭上（以防烤焦）。
2. 用串烤的不鏽鋼針或竹籤，將香魚彎曲定型。
3. 烤箱預熱至 250℃，將魚放入以大火烤，中途翻 2 次面，以使火力平均。
4. 見魚已部分呈焦黃，且魚肉已熟時即可抽出串針，放在碟內，附上檸檬片上桌。

註：除香魚之外，赤鯮魚、嘉鱲魚、金線魚、秋刀魚、四破魚或鯖魚許多魚種均可用此種鹽烤法。

Ingredients

1smelt

Seasonings

1t. wine, 1/2T. salt

Procedure

1. Rinse fish, pat dry. Sprinkle wine and 1/3t. salt over fish and cover the fin and tail with salt to keep from them burned.
2. Shape smelt with a Bar-B-Q skewer.
3. Preheat oven to 250°C, bake fish over high heat. Turn smelt over twice while baking.
4. When smelt becomes a little brown and done, remove skewer and serve with lemon.

* Besides smelt, many another kinds of fish can be baked in this way.

香蒜奶油烤魚頭
Baked Salmon with Garlic

材料
鮭魚頭 1/2 個、大蒜 5 粒、檸檬 1/2 個

調味料
鹽 2/3 茶匙、酒 1 大匙、奶油 2 大匙
粗粒黑胡椒 1/2 茶匙

做法
1. 鮭魚頭洗淨，撒上鹽和酒醃 20 分鐘。大蒜切片。
2. 烤盤中鋪上鋁箔紙，紙上塗約 1 大匙的奶油，放上鮭魚頭（肉面朝上），將大蒜片散放在魚頭上，再撒上黑胡椒粉。
3. 烤箱預熱至 250℃，放入鮭魚頭烤 10 分鐘後翻面再烤，烤至表皮有焦痕即可取出。附檸檬片一起上桌。

Ingredients
1/2 salmon head, 5 cloves garlic, 1/2 lemon

Seasonings
2/3t. salt, 1T. wine, 1/2t. black pepper, 2T. butter

Procedure
1. Rinse fish head. Sprinkle salt and wine on, marinate for 20 minutes. Slice garlic.
2. Place a aluminum foil on baking try, brush 1T. butter , place salmon head on with skin side down. Sprinkle garlic and black pepper on salmon.
3. Preheat oven to 250°C, bake salmon for 10 minutes, turn it over, continue to bake until the skin gets brown. Serve with lemon.

焗烤鮪魚塔
Baked Tuna Fish with Mushroom

材料
新鮮香菇 5 個、罐頭鮪魚 1 罐、芹菜末 1 大匙
麵包粉 2 大匙、奶水 1 大匙、巴西利屑 1 茶匙
蛋黃 1 個

調味料
美奶滋 2 大匙、鹽 1/4 茶匙、胡椒粉 1/6 茶匙
大蒜泥 1 茶匙

做法
1. 將香菇之菇蒂切下，浸泡在鹽水內 1 分鐘，瀝乾並撒少許鹽和太白粉（額外的）在內部。
2. 將鮪魚用叉子壓碎，拌入麵包粉、芹菜碎末、奶水、和調味料，攪拌成糊狀之後，填塞在香菇內部，將表面塗抹光滑並刷上少許蛋黃汁。
3. 在烤盤上鋪 1 張鋁箔紙，排上香菇，移進烤箱內。以 250℃烤 8 ～ 10 分鐘左右，至表面成金黃色便可取出，再撒上巴西利屑。

Ingredients
5 pieces fresh black mushroom, 1 can tuna, 1T. chopped celery, 2T. bread crumbs, 1T. milk, 1t. chopped parsley, 1 egg yolk

Seasonings
2T. mayonnaise, 1/4salt, 1/6t. pepper, 1t. smashed garlic

Procedure
1. Trim mushrooms. Soak in salty water for1 minute, drain and sprinkle some cornstarch on the inside.
2. Smash tuna with a fork, mix with bread crumbs, celery, milk, garlic, mayonnaise, salt and pepper. Stuffed onto mushrooms. Smooth the surface, brush with some egg yolk.
3. Place aluminum foil on baking try, arrange mushrooms on. Bake over 250°C for 8~10 minutes until the surface gets golden brown. Remove and sprinkle with parsley.

烤魚排佐芥末醬汁
Baked Fish Fillet with Mustard Sauce

材料
青衣 1 條（或鱸魚、石斑、鯧魚）、綠蘆筍、紅蕃茄

調味料
Ⓐ鹽 1/3 茶匙、酒 2 茶匙
Ⓑ油 1 大匙、奶油 1 大匙、麵粉 2 大匙、清湯 1 杯、芥末醬 1 大匙
鹽 1/3 茶匙、檸檬汁 1/2 大匙、糖 1 茶匙

做法
1. 青衣由背部下刀，沿著大骨片下兩片魚肉，撒鹽和酒醃 5 分鐘。
2. 平底鍋塗少許油，魚肉面朝下煎 30 秒，微有焦痕，翻面再煎一下。
3. 烤盤上抹奶油，放上魚肉，淋下清湯或水 1/4 杯，再移入預熱至 250℃的烤箱，大火蒸烤 5 分鐘至熟。
4. 鍋中混合油和奶油，炒香麵粉，加入清湯等調味料Ⓑ調勻，做成芥末醬汁搭配魚排和配菜上桌。

Ingredients
1 fish, asparagus, tomato

Seasonings
Ⓐ1/3t.salt, 2t. wine
Ⓑ1T. oil, 1T. butter, 2T. flour, 1C. soup stock, 1T. mustard, 1/3t. salt, 1/2T. lemon juice, 1t. sugar

Procedure
1. Remove head and bones from fish to get 2 pieces of fillet. Sprinkle salt and pepper, stay for 5 minutes.
2. Grease the pan, fry fish fillet with meat side down foe 30 seconds, when it gets lightly brown, turn it over and fry a little while.
3. Grease the baking try with butter, place fish on, pour 1/4C. soup stock or water in. remove to the preheated oven (about 250°C), bake over high heat for about 5 minutes.
4. Stir-fry flour with mixed oil and butter, add seasoningsⒷ, stir evenly to make the mustard sauce. Serve fish with mustard sauce and cooked vegetables.

冷食

Chilled

在「冷食」之前——

對日本人來說，最新鮮、最美味的魚就要生吃。未經烹煮的魚肉口感特別，有的是QQ的，有的是嫩嫩的，帶有油花的魚又有特殊的香氣。台灣現在愛吃生魚片的人也越來越多，頂級的黑鮪每年5月前後來報到，總是掀起一陣旋風。

做生食的菜式當然首重衛生，一定要用切熟食的砧板和刀子切配，用過之後砧板也要洗乾淨，以免放久仍會污染。傳統日式的生魚片一般僅搭配醬油、山葵和白蘿蔔絲上桌。也吃過鮭魚生魚片包捲洋蔥絲，十分美味。

其實用生魚片搭配一些可以生食的蔬菜或水果，再配一個帶有微微果酸的醬汁，是非常健康、爽口的吃法。至於醬汁就可以自由發揮了，日式帶有柚子香的醋汁是現在很流行的，中國式三合油（醬油、麻油、醋）也很合味。義大利菜中也有將鮪魚微微的煎一下，切片後淋一個醬汁，做起來也不難。

另外有一些適合冷食、卻不是生食的魚料理，是和一般魚的吃法不同的。通常魚在冷了之後就不好吃了，魚肉會變硬、有的魚腥味就出來了。但是有些魚則不然，其中江浙式的「蔥㸆鯽魚」、「蘇式燻魚」就非常有名，另外還有「酥魚」、「凍魚」也是宜冷食的。最主要是將魚炸透，再浸泡在味道較濃的汁中，西式的吃法喜歡用橄欖油浸泡，並搭配解油膩又健康的洋蔥、大蒜和新鮮蔬菜。

這些浸泡出來適合冷食的魚料理，最大的好處就是耐放，儲存得當的話，可以放上一個星期，而且炸酥的魚連骨頭一起吃，更可攝取到鈣質，為營養加分呢！

一般推薦可用於冷食的魚種有

沙梭魚・草魚中段・小黃魚・黑鮪魚切片・秋刀魚・鯽魚・四破魚・丁香魚
旗魚切片・鯧魚

燻鮭魚海苔捲
Smoked Salmon Rolls

材料
燻鮭魚 10 片、海苔片 10 片
洋蔥屑 1/2 杯、鮭魚卵適量

調味料
山葵、醬油各適量

做法
1. 取海苔一片，上面放一片燻鮭魚，塗少許山葵，再放上洋蔥屑和鮭魚卵。
2. 捲起來便可食用。也可以少量沾一些醬油。

Ingredients
10pieces smoke salmon, 10 pieces dried seaweed, 1/2C. chopped onion, salmon spawn

Seasonings
mustard and soy sauce

Procedure
1. Place one piece of salmon on top of dried sea weed, put a little mustard on, add some chopped onion and salmon spawn.
2. Roll and eat. You may dip a little of soy sauce while eating.

燻鮭魚小塔
Smoked Salmon Salad

材料
燻鮭魚 150 公克、酸黃瓜屑 2 大匙
土司麵包 6 或 12 片圓形小餅乾
白煮蛋 2 個、洋蔥屑 2 大匙
海苔粉少許

調味汁
美奶滋 2 大匙、黃色芥末醬 1 茶匙
果糖 1 茶匙、檸檬汁 1 茶匙、鹽少許
胡椒粉少許

做法
1. 燻鮭魚切成小片。白煮蛋切碎，和洋蔥屑、酸黃瓜屑一起都放碗中，加所有調味料調勻，再加入燻鮭魚略拌。
2. 土司麵包用小杯子扣出圓片，放上約 1 大匙的燻鮭魚沙拉，再放上一小片燻鮭魚做點綴，撒下海苔粉即可。

Ingredients
150g. smoke salmon, 2T. chopped onion, 2T. chopped pickle, 2 boiled eggs, 6 pieces bread or 12 pieces cookies, seaweed powder

Seasonings
2T. mayonnaise, 1t. syrup, 1t. lemon juice, 1t. yellow mustard, a pinch of salt and pepper

Procedure
1. Dice smoke salmon. Chop boiled eggs. Put together with onion and pickle, mix all seasonings well, add smoke salmon.
2. Use a small cup to mold the bread, put about 1T. smoke salmon salad on bread, decorate with a little of salmon and chopped parsley.

廣式魚生沙拉
Salmon Salad, Cantonese Style

材料

1. 可生食的鮭魚 200 公克
   白芝麻 1 大匙、檸檬皮屑 1/2 茶匙
2. 西洋生菜絲、白蘿蔔絲、胡蘿蔔絲
   黃瓜絲、薑絲、香菜各 2/3 杯
3. 熟腰果或五香花生米、薄脆各 1/2 杯

調味料

醬油 4 大匙、水果醋 2 大匙、糖 1 茶匙
橄欖油 1 大匙、麻油 1 茶匙、胡椒粉少許

做法

1. 將鮭魚生魚片切成 2 寸多長片或切成粗條亦可，排在大盤中央。四周圍放切成細絲的各種材料、香菜和切碎之熟腰果及炸香之薄脆。撒下胡椒粉、檸檬皮屑與白芝麻（炒香略壓碎）。
2. 在一只碗內調勻調味料。上桌後再將汁淋下，當場挑拌均勻取食即可。

註：廣東人把炸過的餛飩皮稱為薄脆，也可以用玉米脆片代替。

Ingredients

200g. salmon, 2/3 cup of each: radish, carrot, cucumber, lettuce, ginger, cilantro, cashew nuts(or peanuts),fried won-ton wrapper, 1/2t. shredded lemon rind, 1T. white sesame seeds

Seasonings

4T. soy sauce, 2T. fruit vinegar, 1t. sugar, 1T. olive oil, 1t. sesame oil, a pinch of pepper

Procedure

1. Slice salmon, arrange on serving plate. Place all shredded ingredients, crushed cashew nuts and fried won-ton wrapper around salmon. Sprinkle pepper , shredded lemon rind ,and sesame seeds(lightly smashed) on top.
2. Mix all seasonings in a small bowl, serve with salmon salad. Pour sauce over just before eating.

*You may use corn flakes instead of fried won-ton wrapper.

和風鮪魚水果沙拉
Tuna and Fruit Salad, Japanese Style

材料
鮪魚生魚片 150 公克、西生菜 1/2 球
炒過的白芝麻 1 大匙
蘋果、芭樂、蓮霧、蘆筍各適量

調味料
柴魚醬油 2 大匙、柚子醋 1 1/2 大匙
味醂 1 大匙

做法
1. 西生菜切絲，泡過冰水後擦乾，排在盤中，撒上白芝麻。
2. 水果削好切成適當大小。蘆筍需先燙過，浸泡冰水後再用。
3. 鮪魚在塗了少許橄欖油的平底鍋上快速煎一下，煎至喜愛的熟度，取出切片排盤。
4. 調好調味料後，裝入小碟中一起上桌沾食。

Ingredients
150g. tuna, 1/2 ball lettuce, 1T. stir-fried sesame seeds, apple, guava, bell fruit, asparagus

Seasonings
2T. Japanese soy sauce, 1½T. fruit vinegar, 1T. mirin

Procedure
1. Shred lettuce, soak with iced water, drain. Place on a plate, sprinkle sesame seeds on top.
2. Cut all fruits in suitable size. Blanch asparagus, drain and soak in iced water, drain again and slice. Arrange on top of lettuce with fruits.
3. Brush a little of olive oil on a pan, fry tuna lightly, remove and slice into thin pieces.
4. Mix seasonings well, serve with tuna salad.

蔥燴鯽魚

Stewed Carps with Green Onion

材料
小鯽魚 8 條、蔥 15 支、麻油 1 茶匙

調味料
Ⓐ酒 2 大匙、醬油 3 大匙、醋 2 大匙
Ⓑ醬油 3 大匙、醋 2 大匙、糖 1 大匙
水 4 杯

做法
1. 鯽魚打理乾淨，擦乾水分，用調味料Ⓐ拌醃 30 分鐘。
2. 燒熱油 1 杯，放下鯽魚大火煎黃兩面，盛出。
3. 蔥洗淨，一切為二，用熱油煎成略焦黃且有香氣，放下鯽魚，加入調味料Ⓑ，大火煮滾後改小火慢慢燒煮。
4. 約 2 小時左右，見湯汁已收乾，淋下約 1 茶匙麻油，搖動鍋子使味道均勻，移入盤中待涼透後便可食用。

Ingredients
8 small gold carp, 15 stalks green onion, 1t sesame oil

Seasonings
Ⓐ2T. wine, 3T. soy sauce, 2T. vinegar
Ⓑ3T. soy sauce, 2T. vinegar, 1T. sugar, 4C. water

Procedure
1. Rinse and pat dry the fish. Marinate with seasoningsⒶ for 30 minutes.
2. Heat 1C. oil to deep-fry fish until brown, remove.
2. Trim green onion, cut into long sections. Fry with hot oil until fragrant and brown. Return fish to wok, add seasoningsⒷ, turn to low heat when it boils.
3. Simmer for about 2 hours, when the liquid is absorbed, drizzle 1T. sesame oil. Shake wok to make the taste evenly. Remove to plate, serve it after cools.

蘇式燻魚
Deep-fried Spicy Fish

材料
草魚 1.5 斤、蔥 3 支、薑 5 片

調味料
Ⓐ醬油 4 大匙、酒 1 大匙
Ⓑ蔥 1 支（切段）、醬油 3 大匙、糖 4 大匙
麻油 1/2 大匙、八角 1 顆、五香粉 1 茶匙
水 1/2 杯

做法
1. 草魚洗淨擦乾，由背部剖開成兩大塊，再橫片成大斜片。蔥、薑拍碎和調味料Ⓐ拌勻，放入魚片醃約 30 分鐘。
2. 用約 1 大匙油爆香蔥段，倒入調味料Ⓑ煮滾。
3. 燒熱炸油，分兩批放入魚片炸酥。當第一批炸好時，立刻泡入調味料Ⓑ中，泡約 3 ～ 4 分鐘。
4. 當第二批魚炸好後，先將第一批魚取出，再泡入第二批，泡 3 ～ 4 分鐘後，加入第一批同泡至涼。

Ingredients
900g. grass carp, 3 stalks green onion, 5 slices ginger

Seasonings
Ⓐ4T. soy sauce, 1T. wine
Ⓑ1 stalk green onion, 3T. soy sauce, 4T. sugar, 1/2C. water, 1t. five spicy powder, 1/2T. sesame oil

Procedure
1. Rinse and halve the grass carp, slice diagonally into large pieces. Crush green onion and ginger, mix with seasoningsⒶ for 30 minutes.
2. Stir-fry green onion with 1T. oil, add seasonings Ⓑ, bring to a boil.
3. Divide fish into two parts, deep-fry them separately until very crispy. Soak into Ⓑ sauce for 3~4 minutes.
4. Remove the first group of fish when the second group is fried, soak the second group for 3~4 minutes. Add first group into sauce, soak until cools. It tastes better when it serve cold.

材料

旗魚生魚片 250 公克、水梨 200 公克、小黃瓜 1 支、生菜葉 2 片
松子 2 大匙、黑白芝麻、香菜、蔥絲各適量

檸檬醋

糖 2 大匙、冷開水 2 大匙、檸檬汁 2 大匙

調味料

辣豆瓣醬 1½ 大匙、糖 3 大匙、蔥末 1½大匙、蒜末 1/2 大匙、薑末 1 茶匙
麻油 1/2 大匙、檸檬汁 1/2 大匙

做法

1. 旗魚肉切成粗條。水梨削皮，切絲。小黃瓜也切絲。
2. 小鍋中將調味料小火煮滾，煮時要多攪動。關火後加入麻油。放涼後再加檸檬汁調勻，做成辣醬。
3. 將魚條和黃瓜絲放入辣醬中拌勻。
4. 梨絲和檸檬醋拌好。
5. 生菜葉墊底，中間放梨絲，再將魚條分別堆放盤中，撒下松子、黑、白芝麻、香菜、蔥絲等，食用前拌勻便可。

Ingredients

250g. mackerel, 200g. pear, 1 cucumber, 2pieces lettuce, 2T. pine nuts, black and white sesame seeds, cilantro, green onion shreds

Lemon vinegar

2T. sugar, 2T. water, 2T. lemon juice

Seasonings

1½T. hot bean paste, 3T. sugar, 1½T. chopped green onion, 1/2T. chopped garlic, 1t. chopped ginger, 1/2T. sesame oil, 1/2T. lemon juice

Procedure

1. Cut mackerel into strips. Peel and shred the pear. Shred cucumber.
2. Bring seasonings to a boil, stir the sauce while cook it. Add sesame oil after turn off the heat, and add lemon juice after it gets cold.
3. Mix fish, cucumber and the sauce together.
4. Mix pear and lemon vinegar.
5. Place lettuce on serving plate, put pear on lettuce, arrange mackerel salad on plate. Sprinkle pine nuts, sesame seeds, cilantro and green onion shreds on top at last.

煙燻鯧魚
Smoked Pomfret

材料
鯧魚 1 條、美奶滋 2 ～ 3 大匙、鋁箔紙 1 張

醃魚料
蔥 2 支、薑 3 片、鹽 1 茶匙、酒 1 大匙

燻料
麵粉 2/3 杯、黃糖 2/3 杯、紅茶葉 2/3 杯

做法
1. 鯧魚斜切成 4 片後用醃魚料醃 20 分鐘。
2. 蒸鍋中水煮開，放入魚蒸約 8 ～ 10 分鐘至熟，取出放涼。
3. 炒鍋中鋪鋁箔紙一張，上面放上燻料，擺上鐵架（需先塗油），把魚排放在架子上，蓋上鍋蓋。
4. 先用中火，待出煙後改成小火，燻 15 分鐘後翻面，再燻約 10 ～ 12 分鐘，見顏色已變為淺褐色，取出放涼，附美奶滋上桌沾食。

Ingredients
1 pomfret, 2~3T. mayonnaise, a piece of aluminum foil

Marinade seasonings
2 stalks green onion, 3 slices ginger, 1t. salt, 1T. wine

Smoking ingredients
2/3C. flour, 2/3C. brown sugar, 2/3C. black tea leaves

Procedure
1. Slice fish into big pieces. Marinate for 20 minutes.
2. Steam fish for about8~10 minutes. Remove and let cools.
3. Place aluminum foil on a wok, place smoking ingredients on foil, put a greased rack on top. Arrange fish on the rack, covered.
4. Smoke over medium heat, reduce to low while the smokes come out. Turn fish over after 15 minutes, smoke for 10~12minutes more or when fish becomes brown enough. Remove and let cools.

咖哩鮮蔬浸魚
Curry Fish with Vegetables

材料
魚肉 300 公克、青椒 1 個、胡蘿蔔 60 公克
西芹 1 支

調味料
Ⓐ醃魚料：鹽 1/2 茶匙、胡椒粉少許
牛奶 1/3 杯
Ⓑ炸魚料：麵粉 1/2 杯、咖哩粉 1 大匙
Ⓒ浸魚料：醋 4 大匙、橄欖油 6 大匙
白酒 1 大匙、鹽 1/2 茶匙
胡椒粉適量

做法
1. 魚肉切斜片，用醃料拌醃 10 分鐘。
2. 西芹切片、青椒切圈、胡蘿蔔切薄片，全部放入盆中，加入浸魚料，油要慢慢地加入，邊加邊攪。
3. 炸魚料先混合，魚片沾粉料後用油炸酥脆，撈出後趁熱和蔬菜料一起在汁中浸泡，十餘分鐘入味後便可食用。

Ingredients
300g. fish fillet, 1 stalk celery, 1 green pepper, 60g. carrot

Seasonings
ⒶTo marinate fish: 1/2t. salt, a pinch of pepper, 1/3C. milk
ⒷTo coat fish: 1/2C. flour, 1T. curry powder
ⒸTo soak fish: 4T. vinegar, 1T. white wine, 1/2t. salt, pepper, 6T. olive oil

Procedure
1. Slice fish, marinate for 10 minutes.
2. Slice celery, green pepper and carrot. Put into a bowl, add the soaking seasonings in, stir slowly and constantly while adding the oil.
3. Mix flour and curry powder, coat fish with the powder. Deep-fry fish until crispy, drain and soak into the olive oil sauce immediately. Soak over 10 minutes.

義大利式酒醋浸魚
Cold Fish, Italian Style

材料

小沙丁魚或四破魚、麵粉 5 大匙
洋蔥 1 個、大蒜 5 粒、西芹 2 支

調味料

Ⓐ鹽 1/2 茶匙、胡椒粉少許、酒 1 大匙
Ⓑ橄欖油 2 大匙、紅酒 1/2 杯、糖 2 大匙
檸檬汁 2 大匙、紅酒醋 1/2 杯

做法

1. 沙丁魚洗淨內臟，用調味料Ⓐ拌勻，醃 10 分鐘。
2. 洋蔥、西芹分別切粗條，大蒜切片，和調味料Ⓑ拌勻。
3. 魚擦乾水分，沾上麵粉，投入熱油中炸至十分酥脆（可以炸 2 次）。趁熱泡入紅酒醋汁中，浸泡 6 小時以上。

Ingredients

small sardine or other small fish, 1 onion, 5 cloves garlic, 2 stalks celery, 5T. flour

Seasonings

Ⓐ1/2t. salt, pepper, 1T. wine
Ⓑ2T. olive oil, 1/2C. red wine, 1/2C. 1/2C. red wine vinegar, 2T. sugar, 2T. lemon juice

Procedure

1. Rinse sardine, mix with seasoningsⒶ, stay for 10 minutes.
2. Cut onion and celery into strips, slice garlic, mix them with seasoningsⒷ.
3. Pat the fish dry, coat with flour. Deep-fry in hot oil until fish becomes very crispy (you may deep-fry twice). Drain and soak in the wine vinegar sauce, soak over 6 hours.

鮮茄油醋汁浸魚
Cold Fish with Tomatoes

材料
中小型魚 4 條、蕃茄 1 個、西芹 10 公分長 1 段
小黃瓜 1 條、小洋蔥 1 個、麵粉 3 大匙

調味料
醋 1 大匙、鹽 1/2 茶匙、橄欖油 4½ 大匙
胡椒粉少許

做法
1. 洋蔥切薄片成圓圈狀。蕃茄、小黃瓜和西芹都切成約 1 公分的小丁。
2. 大碗中放醋、鹽和胡椒粉，慢慢淋下橄欖油，邊淋邊攪拌，最後放入洋蔥圈，完成油醋汁。
3. 魚撒上適量的鹽醃 10 分鐘，拌上麵粉，入熱油中炸至熟且酥（或用少量油煎熟），撈出後立刻泡入油醋汁中，浸泡 1 個小時。
4. 取出魚後，將蔬菜丁放入汁中，泡約 5 分鐘，連汁帶蔬菜料一起淋在魚身上即可。

Ingredients
4 medium sized fish, 1 tomato, 1 cucumber, 1 piece celery(about 10 cm long), 1 small onion, 3T. flour

Seasonings
1T. vinegar, 1/2t. salt, pepper, 4½T. olive oil

Procedure
1. Slice onion. Dice tomato, cucumber and celery into 1cm cubes.
2. Put vinegar, salt and pepper in a large bowl, add olive oil slowly, stir it constantly while adding. Put onion in at last.
3. Sprinkle salt on fish, marinate for about 10 minutes. Cover with flour, deep-fry in the very hot oil until done and brown(or you may fry them with little oil). Drain and soak into the sauce immediately. Soak for at least 1 hour.
4. Remove fish, add tomato, cucumber and celery cubes into the juice, soak for 5 minutes. Pour the juice and vegetables over fish then serve.

小魚花生
Crispy Fish with Peanuts

材料
魩仔魚 200 公克、紅辣椒 2 支、蔥 2 支
油炸花生 1 杯、大蒜碎屑 1 大匙

調味料
鹽、胡椒粉各少許

做法
1. 紅辣椒和蔥切碎。油炸花生米去皮。
2. 魩仔魚洗淨、擦乾水分，放入熱油中炸至酥脆且成金黃色，撈出瀝淨油。
3. 再將大蒜、紅辣椒和蔥花也快速過油一下，全部撈出（或用少量油炒香）。
4. 將乾鍋燒熱，放入所有材料大火拌炒，撒下鹽和胡椒粉兜炒均勻即可，放涼一些後再吃才更加香酥脆。

Ingredients
200g. larval fish, 1C. deep-fried peanuts, 2 red chili, 2 stalks green onion, 1T. chopped garlic

Seasonings
a little of salt and pepper

Procedure
1. Chop red chili and green onion. Peel peanuts.
2. Rinse small fish, pat dry. Deep-fry in hot oil to crispy and light brown. Drain.
3. Run garlic, red chili and green onion through oil, drain.(or you may stir-fry them until fragrant).
4. Heat the wok without oil, stir-fry all the ingredients over high heat, sprinkle salt and pepper in, mix evenly. Serve after it gets cooler.

蜜汁丁香
Sweet & Crispy Fish

材料
丁香魚乾 150 公克、白芝麻 1 茶匙
油 1 杯

調味料
醬油 3 大匙、糖 3½ 大匙、酒 1/2 大匙
薑汁 1/3 茶匙

做法
1. 小魚乾篩去渣子後裝在大碗內，用熱水泡洗一下隨即瀝淨擦乾。
2. 在鍋內燒熱油後，將小魚乾落鍋，用大火炸約 4 分鐘至酥脆，撈出。
3. 另用 1 茶匙油炒醬油、糖、酒和薑汁，炒勻後放入丁香魚快速拌勻，至糖汁濃稠，撒下炒過的芝麻即可。涼後食用較酥脆入味。

Ingredients
150g. dried fish, 1C. oil, 1t. sesame seeds

Seasonings
3T. soy sauce, 3½T. sugar, 1/2T. wine, 1/3t. ginger juice

Procedure
1. Rinse dried fish, soak in hot water for 1 minute, drain and pat dry.
2. Heat 1C. oil to smoking hot, deep-fry the dried fish over high heat until crispy(about 4 minutes), drain.
3. Heat another 1T. oil to stir-fry the seasonings, add fish in, mix evenly until the sauce is very sticky, sprinkle fried sesame seeds on fish. Remove. It will taste better when it gets cold.

Soups 湯

在「煮湯」之前——

常有學生、朋友問我，煮魚湯時，是該煮久一點，使湯的味道濃一點？還是滾一會兒等魚熟了就關火，以保持魚肉的嫩度？的確，我們做家常菜無法像講究的美食，先熬好魚高湯，再用高湯來煮魚湯，也不願意用一些添加物增鮮，要吃魚肉還是喝湯？得要有個取捨。當然，也是有辦法改善的。

就以「蘿蔔絲鮮魚湯」這道簡單的家常湯來說，煮了 20 ～ 25 分鐘，魚肉吃起來仍有鮮味，但湯的的味道卻不夠鮮美，因此可以加一些蛤蜊來增鮮。另外就要利用「爆鍋」來增加香氣。雖然食譜中只是簡單的說：「用油爆香蔥段和薑片，再把魚煎一下，淋酒」。可是這些步驟一定要做徹底，把蔥、薑和魚煎得黃黃的，再烹上酒，以鍋子的熱氣使酒香散發出來，再加水來煮，這樣魚湯就會有香氣，好喝多了。

除了用爆鍋之外，就要像煮紅魚麵一樣，先把魚撈出，剔下魚肉後把魚頭、魚骨和魚鰭等全部去熬煮成魚高湯，這種方法煮來湯鮮、肉嫩、營養又好，最適合老人和孩子們。

魚除了煮清湯外，還可以加重一點的調味料，煮成不同口味的魚湯，味噌湯是最普遍的。另外魚的加工製品——魚丸，也是許多人喜歡的。但是家庭手工製作的魚丸不會有脆脆的口感，多是較滑嫩的。

煮魚湯，在大火煮滾後就要改小火把味道煮出來。雖說用「小火」，但仍要讓湯呈現滾動的狀態，才能把味道煮出來。因為魚沒有用熱水川燙過，而且魚肉中富含蛋白質，因此煮魚湯時難免會有些浮沫，只要用湯勺撇掉就可以了。

煮魚湯時可以加自己喜歡的材料一起煮，像「砂鍋魚頭」就是傳統的江浙名菜，但是每家餐廳都做得不一樣，各有創意。主要是這個煎得香香的魚頭，搭配任何配料都很合，都可以燉煮出濃郁的滋味。「韓國辣魚鍋」是我多年前在洛杉磯吃的韓國鍋，留下深刻印象，這兩年台灣流行韓國泡菜，它加上魚也可以做成火鍋形式，正好可以和大家分享。

一般推薦可用於湯的魚種有

鰱魚頭 ・ 赤鯮魚 ・ 紅條 ・ 黃魚 ・ 虱目魚 ・ 帶魚 ・ 黑毛 ・ 潮鯛魚片

蘿蔔絲鮮魚湯
Fish Soup with Radish and Clams

材料
新鮮紅尼羅魚 1 條、白蘿蔔 450 公克
蛤蜊 300 公克、蔥 2 支、薑 2 片
香菜少許

調味料
酒 1 大匙、水 6 杯、鹽、胡椒粉各適量

做法
1. 魚打理乾淨，白蘿蔔切絲，蔥切長段。
2. 用 2 大匙油煎香蔥段和薑片，再放入魚略煎一下，淋下酒烹香，再加水 6 杯，煮滾後改小火煮 5 分鐘。
3. 加入蘿蔔絲，再煮至蘿蔔絲夠軟，加入蛤蜊煮至開口，加鹽和胡椒粉調味。裝碗後撒下香菜即可。

Ingredients
1 red mouth breeder, 450g. radish, 300g. clams, 2stalks green onion, 2 slices ginger, a little of cilantro

Seasonings
1T. wine, 6C. water, salt and pepper to taste

Procedure
1. Rinse fish. Shred the radish. Cut green onion into sections.
2. Heat 2T. oil to stir-fry green onion and ginger, add fish in, fry until little brown. Sprinkle wine and water, bring to a boil, reduce the heat to low, simmer for 5 minutes.
3. Add radish in, cook until soft enough. Add clams in, cook until done. Seasoning with salt and pepper. Remove to the soup bowl, put some cilantro on top of fish.

醋椒魚湯
Fish Soup with Vinegar and Pepper

材料
赤鯮魚 1 條、蔥 4 支、香菜段 1/2 杯
薑 2 片

調味料
酒 1 大匙、醬油 2 大匙、鹽 1/2 茶匙
胡椒粉 1/4 茶匙、麻油 1/2 茶匙
醋 2 大匙

做法
1. 將魚清洗乾淨，蔥 2 支切段，另 2 支切蔥花。
2. 鍋中燒熱 2 大匙油，放入赤鯮魚煎黃兩面，同時放下蔥段和薑片煎黃，淋下酒及醬油，加入 5 杯水及鹽，煮滾後改用小火燉煮 20 分鐘。
3. 湯碗中放進蔥花、香菜段、醋、胡椒粉及麻油，將煮好的魚湯沖入即可。

Ingredients
1 red sea bream, 4 stalks green onion, 1/2C. cilantro sections, 2 slices ginger

Seasonings
1T. wine, 2T. soy sauce, 1/2t. salt, 2T. vinegar, 1/4t. pepper, 1/2t. sesame oil

Procedure
1. Rinse fish. Cut 2 stalks green onion into sections and chop the other 2 stalks.
2. Heat 2T. oil to fry the fish until both sides get brown, add green onion sections and ginger in, fry until it also gets brown. Sprinkle wine and soy sauce, add water and salt, bring to a boil. Turn to low heat, simmer for 20 minutes.
3. In a large soup bowl, place chopped green onion, cilantro, vinegar, pepper and sesame oil in, pour the cooked fish soup into the bowl.

味噌魚湯
Miso Soup

材料
新鮮紅斑 1 段（約 300 公克）
嫩豆腐 1 塊、蔥粒 3 大匙

調味料
柴魚片 1 小包、味噌 3 大匙

做法
1. 魚洗淨切塊。豆腐也切塊。
2. 在湯鍋內燒滾 6 杯水，放下柴魚片後將火關熄 5 分鐘，撈棄柴魚片。
3. 放下魚塊，用小火煮約 10 分鐘。加入豆腐，續煮 3 分鐘。
4. 將味噌放在小篩網中，再落入湯內，用湯匙磨壓味噌，使其溶解到湯內，嚐過鹹度，如不夠鹹可酌加少許鹽。再煮至沸滾即立即熄火，撒下蔥粒。

Ingredients
1piece of red grouper(about 300g.), 1 piece bean curd, 3T. chopped green onion

Seasonings
1 small pack dried bonito pieces, 3T. miso

Procedure
1. Rinse fish , cut into small pieces. Dice bean curd.
2. Bring 6 cups of water to a boil, add dried bonito pieces in, turn off the heat, soak for 5 minutes, drain off the bonito.
3. Put grouper into the soup. Cook over low heat for about 10 minutes. Add bean curd in, cook for 3 minutes more.
4. Place miso in a small strainer, sink the strainer into the soup, stir miso with a spoon, so the miso will dissolve in soup. Add some salt if the soup is not salty enough. Turn off the heat as soon as the soup is boiled, sprinkle green onion in soup.

越式魚酸湯
Sour Soup, Vietnam Style

材料
鯛魚片 250 公克、罐頭鳳梨 1 小罐
洋芹菜 2 支、檸檬 1/2 個、洋蔥 1/2 個
蕃茄 2 個、高湯 1 罐、九層塔、蔥花

調味料
酸湯粉 1/2 包、魚露 2 大匙、糖 1 茶匙
胡椒粉適量

做法
1. 魚片橫切成大片。蕃茄切塊，鳳梨片切小片，洋芹菜斜切片，洋蔥切絲。
2. 鍋中放高湯和水共計 6 杯，加入蕃茄塊、鳳梨片、洋蔥絲同煮，約 5 分鐘後加洋芹片和湯粉、魚露、糖、胡椒粉調味，再煮一會兒。
3. 放入鯛魚片，並加入 1 大匙油，再煮至滾且魚已熟時滴下檸檬汁，關火裝碗，再放下九層塔和蔥花。

註：越氏酸湯粉又稱答滿林湯粉，是東南亞一帶常用的做湯調味粉料。

Ingredients
250g. fish fillet, 4pieces of pineapple, 2 tomatoes, 2 stalks celery, 1/2 onion, 1 can soup stock, basil, chopped green onion, 1/2 lemon

Seasonings
1/2T. Tamarind Soup Base, 2T. fish paste, 1t. sugar, pepper

Procedure
1. Slice fish fillet. Cut tomato, pineapple, celery into small pieces, shred onion.
2. Together with soup stock and water, boil 6 cups of liquid in a pot, add tomato, pineapple and onion, cook for 5 minutes. Add celery and seasonings, cook for a while.
3. Put fish slices and 1T. oil into the soup, cook until fish is done. Drop lemon juice. Remove to the soup bowl, add basil and green onion.

*Tamarind Soup Base is often used in South east Asia cuisine.

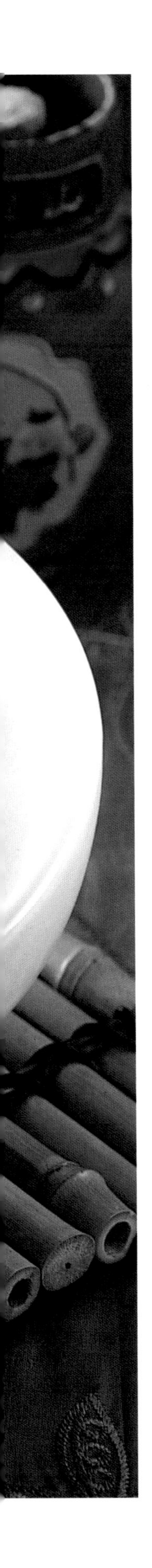

砂鍋魚頭
Fish Head in Casserole

材料
鰱魚頭 1 個、五花肉 3 兩、香菇 6 朵、冬筍 2 支、豆腐 1 塊、白菜 1 斤
粉皮 2 疊、蔥 2 支、薑 2 片、紅辣椒 1 支、青蒜 1 支

調味料
酒 2 大匙、醬油 5 大匙、鹽 1 茶匙、胡椒粉 1/4 茶匙

做法
1. 魚頭先用醬油和酒泡 10 分鐘。五花肉、泡軟的香菇和冬筍都分別切好。青蒜切絲；豆腐切厚片；白菜、粉皮切寬段。
2. 用 5 大匙油將魚頭煎黃，先放入砂鍋中，再把蔥、薑放入鍋中爆香，接著放入五花肉等炒香，淋下剩餘的醬油、辣椒和調味料，注入水 8 杯，大火煮滾，一起倒入砂鍋中，改小火燉煮 1 小時。
3. 放下燙過的白菜和豆腐，再燉煮至白菜夠軟，最後放下切成寬段的粉皮，再煮一滾，撒下青蒜絲即可。

Ingredients
1 silver carp head, 120g. pork(belly part), 6 pieces black mushroom, 2 bamboo shoots, 1 piece bean curd, 600g. Chinese cabbage, 2 pieces mung bean sheet, 2 stalks green onion, 2 slices ginger, 1 red chili, 1 green garlic

Seasonings
2T. wine, 5T. soy sauce, 1t. salt, 1/4t. pepper

Procedure
1. Soak fish head with soy sauce and wine for 10 minutes. Cut pork, soaked black mushrooms and bamboo shoots into pieces. Shred green garlic.Cut bean curd into squares. Cut Chinese cabbage and mung bean sheet into wide strips.
2. Fry fish head with 5 T. oil until the outside get brown, remove to the casserole. Fry green onion and ginger again, when fragrant, add pork, mushroom and bamboo shoot pieces, stir-fry for a while. Add remaining soy sauce, red chili, seasonings and 8 cups of water in, bring to a boil. Pour all ingredients to casserole, cook over low heat for about 1 hour.
3. Add boiled Chinese cabbage and bean curd into casserole, cook until cabbage is soft enough. Add mung bean sheet, bring to a boil, sprinkle shredded green garlic on top, serve.

北方魚丸湯
Home Style Fish Ball Soup

材料
海鰻或新鮮黃魚 450 公克、韭菜 100 公克
薑汁少許

調味料
Ⓐ鹽 1/3 茶匙、水 4 ～ 5 大匙、醬油 2 茶匙
麻油 1 茶匙
Ⓑ胡椒粉少許、麻油 1 茶匙、香菜段少許

做法
1. 將魚肉刮下放入大碗中，加入鹽和水來攪拌魚肉，（水要逐漸加入魚餡中，份量多少視魚新鮮程度而不同，新鮮魚肉可吃進較多的水），至魚肉很有彈性時，加入醬油、薑汁及麻油拌勻。
2. 韭菜洗淨切屑，拌入魚餡中。
3. 鍋中煮滾 5 杯水，加 1/2 茶匙鹽，將魚餡做成丸子，直接投入滾水中川煮，以中火煮至魚丸浮起，撇去表面浮末，再加適量的鹽調味。
4. 大湯碗中加調味料Ⓑ，倒入魚丸湯。

Ingredients
450g. sea eel or yellow croaker, a little ginger juice, 100g. leek

Seasonings
Ⓐ 1/3t. salt, 4~5T. water, 2t. soy sauce, 1t. sesame oil
Ⓑ pepper, 1t. sesame oil, cilantro sections

Procedure
1. Remove all bones, scrape fish finely with the sharp part of a cleaver. Place the fish paste into a large bowl. Stir the paste in one direction with salt and water, add water gradually until sticky. Add soy sauce, ginger juice and sesame oil.
2. Chop leek, mix with the fish paste.
3. Bring 5 C. water to a boil, add 1/2t. salt. Make fish balls with the paste, put fish balls into the soup directly. Cook over medium heat until all balls flow up. Season with more salt.
4. Add seasonings Ⓑ to the soup bowl, pour soup in, serve hot.

湯泡魚生

Sliced Fish Soup, Hunan Style

材料

鯛魚片 250 公克、酒 1 茶匙、生菜絲 1 杯
油條 2 支、蔥屑 2 大匙、白芝麻 1 大匙

調味料

醋 1 茶匙、麻油 1 茶匙、胡椒粉少許
高湯 6 杯、鹽 1½ 茶匙

做法

1. 把魚肉片切成 4 公分寬的薄片，全部放入碗中，淋下少許酒拌勻，醃片刻。
2. 油條切成薄片，入烤箱中烤脆。
3. 生菜切絲後，放入大湯碗中，再放下油條，最後把魚肉鋪在上面。
4. 白芝麻炒香，盛出後稍微壓碎一點，以使香氣透出，撒在魚片上，再撒下蔥屑，並加入胡椒粉、鎮江醋及麻油。
5. 另用一碗盛裝滾燙的高湯（加鹽調味）。兩碗一起上桌，將湯快速倒入魚肉碗中，再將魚肉迅速攪散開，見魚肉變白色已燙熟，便可分裝小碗。

Ingredients

250g. fish fillet, 1t. wine, 2 stalks yu-tiao, 1C. shredded lettuce, 2T. chopped green onion, 1T. sesame seeds

Seasonings

1t. vinegar, 1t. sesame oil, a pinch of pepper, 6C. soup stock, 1½t. salt

Procedure

1. Slice fish into 4cm thin slices, place in a large bowl, mix with a little of wine. Stay for a while.
2. Slice yu-tiao, bake in oven until crispy.
3. Place shredded lettuce into a soup bowl. Add yu-tiao in, arrange fish on top of yu-tiao.
4. Crush the stir-fried sesame seeds slightly, sprinkle on top fish, add green onion, pepper, vinegar and sesame oil in the bowl.
5. Bring soup stock to a boil, serve with the fish fillet. Pour the soup stock over fish, mix quickly, when the color of fish is changed, it is ready for eat.

翡翠魚羹
Fish Potage with Green Vegetable

材料
白色魚肉 250 公克、莧菜 200 公克、筍 1/2 支
營養豆腐 1/2 盒、蛋白 1 個、蔥 2 支、薑 2 片

調味料
Ⓐ鹽 1/4 茶匙、水 2 大匙、蛋白 1/2 大匙
太白粉 2 茶匙。
Ⓑ酒 1/2 大匙、鹽 1 茶匙、太白粉水 1½ 大匙
胡椒粉少許、麻油少許

做法
1. 魚肉切成指甲大小厚片，依序加調味料Ⓐ拌勻，入冰箱冷藏醃 20 分鐘。
2. 莧菜摘好，入滾水中燙煮 10 秒撈出，沖涼，擠乾水分再剁碎，筍煮熟切小片。蔥切段；豆腐切丁。
3. 用 2 大匙熱油煎香蔥段、薑片，淋下酒及水 5 杯煮滾，放下筍片、莧菜和豆腐小丁，再放下魚片攪散並加鹽調味。
4. 煮滾後勾芡，淋下打散的蛋白，撒下胡椒粉、滴少許麻油即可裝碗。

Ingredients
250g. fish fillet, 200g. green vegetable (such as Chinese white spinach or spinach), 1/2 bamboo shoot, 1/2 box bean curd, 2T. egg white, 2 stalks green onion, 2 slices ginger

Seasonings
Ⓐ1/4t. salt, 2T. water, 1/2T. egg white, 2t. cornstarch
Ⓑ1/2T. wine, 1t. salt, 1½T. cornstarch paste, pepper and sesame oil

Procedure
1. Dice fish fillet, marinate with seasonings Ⓐ, store in refrigerator for 20 minutes.
2. Trim spinach, blanch for 10 seconds. Drain and rinse with cold water. Squeeze out the excess water, chop finely. Cook bamboo shoot and slice into small pieces. Cut green onion into sections. Dice bean curd.
3. Fry green onion sections and ginger until brown, sprinkle wine and 5 cups of water, bring to a boil, add bamboo shoot, vegetable, and bean curd, when boil again, add fish in, season with salt.
4. Thicken the soup with cornstarch paste, pour beaten egg white in, add pepper and drizzle sesame oil. Serve.

紅魚湯麵
Noodle with Red Sea Bream

材料
紅魚（赤鯮魚）2 小條、四季豆 10 支、蔥 2 支
細麵 200 公克、薑 2 片

調味料
酒 1 大匙、鹽 1 茶匙、胡椒粉少許、水 5 杯

做法
1. 蔥切段。紅魚刮淨細鱗，擦乾水分，用 2 大匙油煎黃兩面，放入蔥段和薑片同煎香，加入調味料，煮滾後改小火煮 10 分鐘。
2. 撈出紅魚，仔細取下兩邊的大塊魚肉，魚頭、魚骨等零碎部分放回湯中再熬煮 20 ～ 30 分鐘，過濾出湯汁。
3. 四季豆切斜段，放入湯汁中煮。細麵先在滾水中煮 1 ～ 2 分鐘，再一起放入魚湯中煨煮至喜愛的爛度，放回魚肉煮熱，再稍加調味即可。

Ingredients
2 small red sea bream, 10 pieces string bean, 200g. needles, 2 stalks green onion, 2 slices ginger

Seasonings
1T. wine, 1t. salt, a pinch of pepper, 5C. water

Procedure
1. Cut green onion into sections.Rinse fish, pat dry. Fry fish with 2T. oil until brown. Add green onion and ginger to fry, when fragrant, add seasonings. Bring to a boil, cook over low heat for 10 minutes.
2. Remove fish from soup. Remove head and bones, keep the fish fillet as big as possible. Return head and all bones back to soup, simmer for 20~30 minutes. Drain.
3. Slice string beans, put it to the soup. Boil needles in water for 2~3 minutes, remove to soup stock, cook until tender. Put fish back to soup to reheat it, season again with salt.

酸辣魚羹
Sour and Spicy Fish Potage

材料

新鮮魚肉 200 公克、紅蔥酥 2 大匙、香菜少許
大白菜絲、香菇絲、胡蘿蔔絲、金針菇各適量
大蒜泥 2 茶匙、蕃薯粉 1/2 杯

調味料

Ⓐ醬油 1/2 大匙、鹽 1/4 茶匙、五香粉少許
蛋黃 1 個、酒 1/2 大匙
Ⓑ醬油 1 大匙、鹽 1 茶匙、太白粉水 1 大匙
Ⓒ麻油 1 茶匙、胡椒粉 1/4 茶匙、烏醋 2 大匙

做法

1. 魚肉切粗條，用調味料Ⓐ拌勻，醃 10 分鐘。沾上蕃薯粉，用油炸至熟且酥，撈出。
2. 起油鍋用 2 大匙油炒香香菇絲，加入其它蔬菜料和 5 杯水，煮 5 分鐘後加入調味料Ⓑ，再放下魚條和紅蔥酥一滾即關火。
3. 加入調味料Ⓒ，裝入大碗中，視個人喜好可再加入香菜和大蒜泥各酌量。

Ingredients

200g. fish fillet, shredded Chinese cabbage, soaked black mushroom, carrot, needle mushroom, 2T. fried red shallot, cilantro, 2t. smashed garlic, 1/2C. sweet potato powder

Seasonings

Ⓐ1/2T. soy sauce, 1/4t. salt, a pinch of five spicy powder, 1 egg yolk, 1/2T. wine
Ⓑ1T. soy sauce, 1t. salt, 1T. cornstarch paste
Ⓒ1T. sesame oil, 1/4t. pepper, 2T. black vinegar

Procedure

1. Cut fish in big strips. Marinate with seasonings Ⓐ for 10 minutes. Coat with sweet potato powder. Deep-fry with hot oil until done and crispy. Drain.
2. Stir-fry shredded black mushroom with 2T. oil, add other vegetables and 5 cups of water, cook for 5 minutes after it boiled. Add seasonings Ⓑ, fish strips and red shallot, bring to a boil again, turn off the heat.
3. Add seasoningsⒸ, remove to the soup bowl, add cilantro sections and smashed garlic according to personal taste.

西洋菜煲生魚
Fish Soup, Cantonese Style

材料
生魚（鱺魚）1 條、蜜棗 1 顆、薑 4 片
西洋菜 2 把

調味料
酒 1 大匙、鹽適量、白胡椒粉適量

做法
1. 生魚洗淨，剁成 3 ～ 4 段，用滾水川燙 20 ～ 30 秒鐘，取出再洗淨。
2. 西洋菜切成 3 公分長段。
3. 起油鍋用 2 大匙油慢慢煸香薑片，淋下酒隨即倒入水 6 杯和蜜棗，煮滾後放入魚和西洋菜，用小火煲煮 1小時。
4. 加鹽和胡椒粉調味即可。

Ingredients
1 fresh-water goby, 1 grown date, 2 bundles water cress, 4 slices ginger

Seasonings
1 T. wine, salt, white pepper

Procedure
1. Rinse fish. Cut into 3~4 sections. Blanch for 20~30 seconds, rinse again.
2. Cut water cress into 3cm sections.
3. Fry ginger slices with 2T. oil until fragrant. Sprinkle wine, pour 6 cups of water in, add brown date, bring to a boil. Add fish and water cress in, simmer for one hour.
4. Season with salt and pepper.

韓國辣魚鍋
Fish Pot, Korean Style

材料
黃魚 1 條、韓國泡菜 1 杯、白蘿蔔 1/3 條、豆腐 1 塊、茼蒿菜 150 公克
蔥 3 ～ 4 支

調味料
麻油 2 大匙、酒 1 大匙、清湯 5 杯、韓國紅辣醬 1 ～ 2 大匙、鹽 1 茶匙

做法
1. 魚打理乾淨，切成段。蘿蔔切厚片（或者用泡菜中的白蘿蔔）；豆腐切厚片；茼蒿菜洗淨；蔥切段。
2. 砂鍋或厚鐵鍋中用麻油 2 大匙炒香蔥段和魚塊，淋下酒和清湯（包括泡菜的湯汁），同時放下豆腐、白蘿蔔和泡菜同煮，約煮 10 分鐘。
3. 加紅辣醬和鹽調味，最後放下茼蒿菜一滾便關火。整鍋上桌，可當作火鍋，邊吃邊加材料同煮。

Ingredients
1 yellow croaker, 1C. Korean pickle, 1/3 radish, 1 piece bean curd, 150g. green vegetable, 3~4 stalks green onion

Seasonings
2T. sesame oil, 1T. wine, 5c. soup stock, 1~2T. Korean hot red paste, 1t. salt

Procedure
1. Rinse fish, cut into pieces. Slice radish into thick pieces(or you may use the radish from Korean pickle). Cut bean curd into thick pieces. Trim vegetable. Cut green onion into sections.
2. Heat 2T. sesame oil in a casserole or a thick iron pot, stir-fry green onion and fish, sprinkle wine and soup stock(including the juice from pickle). Add bean curd, radish and pickle in, cook for 10 minutes.
3. Add hot red paste and salt to taste, add vegetable, bring to a boil, Turn off the heat. You may serve it as a hot pot dish, put a small stove on table, cook other ingredients while eating.

在家輕鬆做系列

在家燒一手好菜
輕鬆當大廚，天天換菜色！
程安琪　著／169元
用不同食材「燒」出色香味俱全的好菜，安琪老師的50道精選菜肴，讓你輕輕鬆鬆變大廚！

在家燉煮一鍋美味，
47道砂鍋與火鍋料理
程安琪　著／169元
暖呼呼的砂鍋菜、料多味美的火鍋料理，47道安琪老師的精選，讓你善用食材特性，學會鍋的妙用，輕輕鬆鬆燉煮一鍋美味！

在家做醃菜&泡菜
程安琪　著／169元
本書分9大類，從基礎的泡菜製作到多變化醃菜料理，韓國泡菜、東北酸白菜、雪裡紅等，每道菜的烹調步驟清楚，讓你輕鬆就能在家做醃菜與泡菜。

品味生活系列

營養師設計的82道
洗腎保健食譜
衛生福利部桃園醫院營養科　著
楊志雄　攝影／380元
桃醫營養師團隊為洗腎朋友量身打造！內容兼顧葷食&素食者，字體舒適易讀、作法簡單好上手照著食譜做，洗腎朋友也可以輕鬆品嘗美食！

燉一鍋×幸福
愛蜜莉　著／365元
因為意外遇見一只鑄鐵鍋，從此愛上料理的愛蜜莉繼《遇見一只鍋》之後第二本廚房手札。書中除了收錄她的私房好菜，還有許多有趣的廚房料理遊戲和心情故事。

健康氣炸鍋
教你做出五星級各國料理
陳秉文　著／楊志雄　攝影／300元
煮父母&單身新貴的料理救星！60道學到賺到的五星級氣炸鍋料理食譜，減油80%，效率UP！健康氣炸鍋的神奇料理術美味零負擔的各國星級料理輕鬆上桌！

首爾糕點主廚的人氣餅乾
卞京煥　著／280元
焦糖杏仁餅乾、紅茶奶油酥餅、摩卡馬卡龍……，超過300張的清楚步驟圖及解說，按照主廚的步驟step by step，你也可以變身糕點達人！

嬰兒副食品聖經
趙素濚　著／600元
最具公信力的小兒科醫生＋超級龜毛的媽媽同時掛保證，最詳盡的嬰幼兒飲食知識、營養美味的副食品，205道精心食譜＋900張超詳細步驟圖，照著本書做寶寶健康又聰明！